Sustainable Development Goals Series

The **Sustainable Development Goals Series** is Springer Nature's inaugural cross-imprint book series that addresses and supports the United Nations' seventeen Sustainable Development Goals. The series fosters comprehensive research focused on these global targets and endeavours to address some of society's greatest grand challenges. The SDGs are inherently multidisciplinary, and they bring people working across different fields together and working towards a common goal. In this spirit, the Sustainable Development Goals series is the first at Springer Nature to publish books under both the Springer and Palgrave Macmillan imprints, bringing the strengths of our imprints together.

The Sustainable Development Goals Series is organized into eighteen subseries: one subseries based around each of the seventeen respective Sustainable Development Goals, and an eighteenth subseries, "Connecting the Goals," which serves as a home for volumes addressing multiple goals or studying the SDGs as a whole. Each subseries is guided by an expert Subseries Advisor with years or decades of experience studying and addressing core components of their respective Goal.

The SDG Series has a remit as broad as the SDGs themselves, and contributions are welcome from scientists, academics, policymakers, and researchers working in fields related to any of the seventeen goals. If you are interested in contributing a monograph or curated volume to the series, please contact the Publishers: Zachary Romano [Springer; zachary.romano@springer.com] and Rachael Ballard [Palgrave Macmillan; rachael.ballard@palgrave.com].

More information about this series at http://www.springer.com/series/15486

Andrew Simmons

Climate Change Adaptation Framework and Youth Entrepreneurship in West Africa

Springer

Andrew Simmons
Kingshill Development Solutions (KDS)
Enfield Lock, Middlesex, UK

ISSN 2523-3084 ISSN 2523-3092 (electronic)
Sustainable Development Goals Series
ISBN 978-3-030-85756-1 ISBN 978-3-030-85754-7 (eBook)
https://doi.org/10.1007/978-3-030-85754-7

This Springer imprint is published by the registered company Springer Nature Switzerland AG
The registered company address is: Gewerbestrasse 11, 6330 Cham, Switzerland

This book is dedicated to Ottis Joslyn, climate change specialist, environmental activist, my climate change soldier and friend who passed away earlier this year. He was the former science officer of the Caribbean Community Climate Change Centre (5Cs). Your work and memory will always be with us.

Foreword

The global uneven economic growth has had an impact on the well-being of livelihood and employment opportunities. The continent needs to cope with the ever-increasing depletion of its natural resources, climate change, population growth, and the heightened urgency to create the development of the livelihood enabling environment in a sustained manner.

The non-inclusive trend of Africa's growth and development models is highly dominated by capital intensive sectors that grant little or no benefit to the broader portion of the society that possesses a vast potential in labour force supply. The current economic reality of Africa indicates that it is not able to increase the capacity of its human capital through the development of employment opportunities specifically for marginalized youth and women. The youth segment of the population is being challenged with many issues that require economic, political, and social supports.

This notion indicates that employment opportunities are not growing fast to cope with the ever-increasing youth population worldwide. In the context of African countries, including Sierra Leone, unemployment figures downplay the issue of underemployment and poverty that prevail on a grander scale among the youthful population, more specifically with the rural youth because of meagre employment opportunities. As a result of this, the rural youth usually migrate to urban areas searching for jobs and livelihood opportunities. This situation magnifies the need to create sustainable employment opportunities for African youth. In this regard, the importance of addressing unemployment from an entrepreneurship perspective is the yet untapped potential of many nations of Africa. Cultivating this stronghold will also have its direct contribution to local economic development endeavours.

Dr Andrew Simmons has helped UNDP's Youth Employment and Empowerment Programme in Sierra Leone by playing a dominant role in the formulation of the Blueprint for Sierra Leone Five Year Youth Development Programme that has laid the strategic youth employment and empowerment areas of the country. He has also helped in designing the Youth Development Programmes Coordination Framework, which was aimed at capacitating concerned governmental institutions such as the Ministry of Youth (MOYA) and the National Youth Commission (NAYCOM) of Sierra Leone. Dr Andrew Simmons has also assisted in documenting best practices and lessons learned from YEEP to be used as a knowledge platform in enhancing youth development ventures. In addition to that, in his role to lead the evaluation of YEEP, he came up with a body of knowledge on Youth Employment and

Entrepreneurship initiatives to address youth and sustainable livelihood in Africa.

Grounded with this diverse exposure, competencies and knowledge, this body of work has a valuable form of learning to document the processes and frameworks, leading to the successful implementation of youth-focused initiatives, the Youth Employment and Entrepreneurship Project. The book is useful and timely as it focuses on youth entrepreneurship as a form of resilience building in face of a multiplicity of threat, chief among these being climate change that exacerbates youth employment. It brings together top-down and bottom-up approaches to address youth employment and climate change challenges through an entrepreneurship framework in Sierra Leone and beyond.

Many thanks to Dr Andrew Simmons for his wealth of profound knowledge and seasoned professionalism, including his magnificent local, regional, and global roles in combating climate change, enhancing youth development, and capacity development in sustainable development arenas. I want to wish you all the best on your new book entitled *Climate Change Adaptation Framework and Youth Entrepreneurship in Africa*. I am sure it will be most welcomed by policymakers and practitioners alike and will contribute significantly to the body of knowledge on youth development, entrepreneurship, and climate change adaptation.

International Development Expert
Addis Ababa, Ethiopia

Molla Makonnen Alemu

Definition of Terms

Adaptive Capacity – the ability of a system to adjust to climate change, including climate variability and/or extremes, moderate potential damages or cope with consequences, and take advantage of the opportunities climate change may bring and/or cope with the consequences (IPCC, 2007).

Bottom-Up Approach – the approach assumes that if one can address actual vulnerability today, one inevitably reduces future (expected) vulnerability (Burton *et al.* 2002). The bottom-up approach develops a process and procedures for implementation of the approach by identifying the network of actors and stakeholders who are involved in the delivery of services at the local area level by focusing on their goals, strategies, activities and contacts (Hanf, Hjem and Porter, 1978, Stochowiak et al., 2016).

Climate Change – a change of climate which is attributed directly or indirectly to human activity that alters the composition of the global atmosphere and which is in addition to natural climate variability observed over a comparable period (UNFPCCC, 2012 in IPCC, 2014pg.5).

Climate Change Adaptation – refers to the adaptation human driven adjustment in ecological, social or economic systems or policy processes, in response to actual or expected climate stimuli and their effects or impacts (LEG, 2011). Various types of adaptation can be distinguished, including anticipatory and reactive adaptation, private and public adaptation, and autonomous and planned adaptation (IPCC AR4, 2007).

Climate Change Education for Sustainable Development (CCESD) – "Preparing people from all walks of life to plan for, cope with and find solutions for issues that threaten the sustainability of our planet" (UNESCO, 2014 pg. 16).

Empowerment – creating and supporting the enabling conditions under which young people can act on their own behalf, and on their own terms, rather than at the direction of others (Commonwealth Secretariat, 2007)

Exposure – the presence of people, livelihoods, species or ecosystems, environmental functions, services, infrastructure or economies, social or cultural assets in place and settings that could be adversely affected (IPCC, 2001, IPCC, 2014).

Green Jobs – any decent job that contributes to preventing or restoring the quality of the environment, be it in agriculture, industry, services or administration (UNEP, ILO, IOE and ITUC 2008 in Poschen 2015)

A Hybrid Approach to Adaptation – a combination of the micro-level variables of bottom-up and the macro-level variables of the top-down approaches in the development and implementation of climate adaptation projects to benefit from the advantages and strengths of both approaches by enabling the various levels of stakeholders to interact and influence project outcomes (Matland 1995; OECD, 2013). Combining both approaches highlights their strengths and minimizes their weaknesses.

Human Capital – the skills, knowledge, ability to labour and good health enable people to pursue different livelihoods strategies and achieve their livelihoods objectives (Angelsen et al., 2011).

Human Capital Models – presume that an investment in education is made with the expectancy of positive returns to the individual or state on their investment.

Human Capital Theory – theoretical framework responsible for adapting education and development policies (Olaniyan & Okemakinde, 2008).

Life Learning Approach – emphasizes the need for policymakers and education and training institutions to link training and skills provided with market needs as an initial strategy to prepare youth to make full use of opportunities created by climate change in the future (UNDP, 2015).

Livelihood – the capabilities, assets and activities required for a means of living (Chambers & Conway, 1988).

Resilience – the ability of a social or ecological system to absorb disturbances while retaining the same basic structure and ways of functioning, the capacity for self-organization, and the capacity to adapt to stress and change (IPCC AR4, 2007).

Sample – the subgroup of a target population studied, i.e. the group from which data are collected (Angelsen et al., 2011).

Sensitivity – the degree to which a system is affected by or is responsive to climate inducements (Lomos, 2001), i.e. it is also referred to as the degree to which a structure or organism is affected unfavourably or beneficially by climate-related impacts (Lomos, 2001).

Shocks – destructive events that occur unexpectedly and last for noticeably short periods, includes tsunamis, earthquakes, storms, volcanic eruptions, landslides, avalanches and wildfires. (Calgaro et al., 2014).

Small Island Developing States (SIDS) – recognized as a distinct, diverse group of developing countries encountering specific social, economic and environmental vulnerabilities at the United Nations Conference on Environment and Development (UNCED in Rio in 1992), also known as the Earth Summit, held in Rio de Janeiro, Brazil (3–14 June 1992) http://unohrlls.org/about-sids/

Stressors – events that happen slowly with their impacts being felt for longer periods (Calgaro, 2011; Calgaro et al., 2014).

Sustainable Development – "development that meets the needs of the present without compromising the ability of future generations to meet their needs" (United Nations, 1987 in UNESCO, 2014 pg20).

Sustainable Livelihoods – a livelihood is sustainable when it can cope with and recover from the stresses and shocks and maintain or enhance its capabilities and assets both now and in the future without undermining the natural resource base (Chambers & Conway, 1992).

SWOT Analysis – an analytical tool that identifies the internal strengths and weaknesses and external opportunities and threats of an organization or project (Johnson and Scholes, 1999).

Target Population – the group being studied, i.e. the group about which one wishes to draw data for conclusions (Angelsen et al., 2011).

Top-Down Adaptation Approach – planned coordination of intention and action to achieve CCA outcomes imposed by the central authority such as the government (Matland 1995; OECD, 2013). Top-down CCA is more applicable to government agencies.

Vulnerability – the degree to which a system is susceptible to, or unable to cope with adverse effects of climate change, including climate variability and extremes. Vulnerability is a function of the character, magnitude and rate of climate change variation to which a system is exposed, its sensitivity, and its adaptive capacity. Therefore, adaptation would also include any efforts to address these components (IPCC AR4, 2007).

Youth Empowerment – Youth are empowered when they acknowledged that they have or can create choices in life, are aware of the implications of those choices, make informed decisions freely, take action based on that decision and accept responsibility for the consequences of those actions (Commonwealth Secretariat, 2007).

Youth – a cohort of persons from 15 to 24 years of age (UN) or 15 to 29 years (the Commonwealth), (Simmons, 2006). Qualitatively, youth is defined as a phase when a person moves from a time of dependency (childhood) to independence (adulthood) (World Bank, 2007).

Acknowledgement

I want to acknowledge the many people who contributed to the successful completion of this project. Special thanks to Commissioner Thomas Ngolo Kata; Momodu Wudie, Project Manager; Yusuf Kamara, M & M&E Manager; Josep Kpakiwa; and other team members of NAYCOM for their leadership and technical support in ensuring the completion of the project. Special thanks are extended to Felicitas Chi Cho Ata Atanga, Project Manager, and Mary Nguanya Forday and Alex Yeaney, Task Managers, the African Development Bank, for providing consistent technical guidance and finance for the project. To Gianni Bangalore, Head AVSI, Sierra Leone; Pierre Berman, Manager; and Giorgio Cardone, Director of LA Int. Corp. of the Consortium for providing technical support to the YEEP staff to implement the project in Sierra Leone.

Most importantly, I want to recognize the tremendous commitment of the local project team working in collaboration with partner institutions, including universities, colleges and TVET centres, governmental agencies, civil society, and the private sector agencies to implement the project. Special thanks to Allan Que, Key Expert (KE) for Component 1: Skills for Employment and Non-Key Experts (NKE), Joseph Juana (Aquaculture), Memuna Kargbo (Sustainable Tourism) and Alina Kollan (Sustainable Construction). Patrick Amara, Key Expert, and Mary Lincon, Non-Key Expert for Component 2, worked on the Career Guidance and Job Readiness Skills. Richard Yeboah, Key Expert, and Peter Lan, Non-Key Expert of Component 3, worked on the Graduate Entrepreneurship Project. I also want to take the opportunity to recognize the unique contribution of Georgiana Bangura, Jusufu Koroma and Amadu for providing the backup support to keep the office functioning.

Special thanks to the heads of all partner agencies, instructors and counsellors for their commitment to work with the Consortium team in Freetown, Sierra Leone, thus ensuring the successful implementation of the project. I also want to identify the exceptional contribution of Georgiana Bangura, Jusufu Koroma, Amadu Wurie Bah, Yaya Sawaneh and Rebecca Langley for providing administrative, accounting and logistical support. Special thanks to Anoria Simmons, my wife, and my daughter Andree-Ann Simmons, who assisted with the editing and formatting of the book.

Abstract

Over the past decade, Sierra Leone has experienced a series of development challenges relating to its socio-economic vulnerability, including internal conflict and the Ebola epidemic in most recent times. The dire unemployment situation in Sierra Leone is compounded by the emergence of climate change that negatively impacts all aspects of daily life and well-being. Another fundamental challenge facing the country is that the education and training provided do not prepare young people adequately for the job market. Therefore, building youth resilience to adapt to climate change and development calamities is a significant development challenge faced by policymakers in Sierra Leone. These development challenges negatively impact youth and their communities, their enterprises, and their overall well-being, especially in rural and urban centres.

Through the National Youth Commission (NAYCOM), Sierra Leone implemented a Youth Employment and Entrepreneurship Project (YEEP) utilizing the Climate Change Adaptation Framework for Youth Development as the mechanism for driving the project. The YEEP aimed to reinforce the employability of Sierra Leone Youth and enhance their entrepreneurship and employment skills through the development of a framework that integrates education and vocational institutions with job market demand and supply. YEEP linked to the National Medium-Term Strategy of the Government of Sierra Leone, emphasizing Youth Entrepreneurship and Human Capital Development. Youth are encouraged to learn skills to enable them to find and or create their employment. The outcome of the approach has strengthened the occupational levels in the youth population segments of the country and triggered sustained economic development in the growth sectors by establishing youth-led small-scale enterprises. The outcomes of the project include:

- Curricula approved by the National Council for Technical Vocational Education Award Agency of Sierra Leone (NCTVA) in Sustainable Construction (Masonry, Carpentry & Aluminium fabrication), Sustainable Hospitalities (Hotel and Tourism Management), Sustainable Agriculture (Aquaculture and Fisheries Management), Career Guidance and Job Readiness Skills and Entrepreneurship Development (Graduate Entrepreneurship Programme).
- An e-learning mechanism that successfully delivered training and technical support to over 6500 youth in urban and rural areas in Sierra Leone. The training was delivered despite poor Internet connectivity, the absence

of appropriate technologies and youth having no experience in using e-learning during the COVID-19 pandemic.
- Memorandum of understandings (MOUs) with the Government of Sierra Leone and 12 partner institutions to enhance their commitment to implementing the project.
- Change management tools that assess youth development institutions' weaknesses and devising strategies to enhance organizations' institutional capacity to deliver youth development programmes.

This book covers the implementation of the conceptual framework on applying the Climate Change Adaptation Framework for Youth Entrepreneurship in Sierra Lone, West Africa, policy documents on the Public-Private Partnership (PPP) and the YEEP pilots model framework. PPP outlines the policy mechanism for the government to work with the business sector within the TVET education subsector. There are templates on questionnaires, focus group discussions (FGD) with youth and interviews with policymakers and senior managers. It also covers the threats of climate change on youth and the adaptation of a similar process to be used by governments and development organizations in implementing similar youth climate change and sustainable livelihoods projects to build the resilience of youth to climate change in countries in West Africa region.

Contents

Acronyms

AfDB	African Development Bank
AFFORD	African Foundation for Rural Development
AfP	Agenda for Prosperity
AfL	Assessment for Learning
BDS	Business Development Centres
CAPS	Career Advisory and Placement Services
CBA	Competency Based Assessment
CCA	Climate Change Adaptation
CCESD	Climate Change Education for Sustainable Development
CGJRS	Career Guidance and Job Readiness Skills
CEPAD	Community Empowerment for Peace and Agricultural Development
FGD	Focus Group Discussion
GEP	Graduate Entrepreneurship Project
GTI	Government Technical Institute of Kenema
GIZ	German International Cooperation
GYN	Global Youth Network
ILO	International Labour OrganisationOrganization
IPCC	International Panel for Climate Change
ITC	Information and Telecommunication
IT	Information Technology
KE	Key Experts
M&E	Monitoring and Evaluation
MMCET	Milton Margai College of Education and Technology
MOYA	Ministry of Youth Affairs
NAYCOM-SL	National Youth Commission of Sierra Leone
NCTVA	National Council for Technical and Vocational Awards
NKE	Non-Key Expert
NU	Njala University
PI	Partner Institution
PPP	Public-Private Partnership
PLA	Participatory Learning Approach
TAC	Teacher's Advance Certificate
TC	Teachers Certificate
TEC	Tertiary Education Commission
TOT	Training of Trainers
TTD	Teacher Training Department

UN	United Nations
UNDP	United Nations Development Programme
UNEP	United Nations Environment Programmes
UNESCO	United Nations Educational, Scientific and Cultural Organization
YEEP	Youth Employment and Entrepreneurship Project

Enhancing the Capacity of Youth to Adapt to the Impact of Climate Change and Other Development Calamities in Africa

1

Abstract

In this chapter starts broadly by discussing the causes of climate change and how it impacts developing countries globally, specifically young people and their communities. It discusses the outcomes from the 2015 Paris Climate Change agreement and challenges to keep global warming within 1.5–2°C; the problems associated with the impact of climate change on developing countries in Africa, and the relevance of developing climate change adaptation models as a mechanism for achieving sustainable development goals. It highlights the unique challenges associated with the impact of climate change in West Africa.

The chapter then examines the processes relating to the development of the CCA framework for Youth in SIDS in addressing fundamental development challenges posed by the impact of climate change and the necessity for building resilience among youth to adapt to the impact of climate change and identifies the steps in the implementation of the framework. It introduces the 'Youth Employment and Entrepreneurship' project implemented by the Government of Sierra Leone as a mechanism for youth employment through entrepreneurship. It explores the component of the Youth Employment and Entrepreneurship Project (YEEP), the strategies implemented to enhance the efficiency of the institution's capacity of the National Youth Commission to implement YEEP and sets out the aim and objectives of YEEP as a vehicle for addressing employment through training in skills for employment and job readiness, career guidance and entrepreneurship development for Youth in West Africa.

Keywords

Climate change · Sustainable development · Youth development · Entrepreneurship · Employment · Climate change adaptation · Africa · Sierra Leone · Biodiversity · Technical vocational education training · Conceptual framework

Global Climate Change and Biodiversity Challenges

The world fails to grasp the extent of the threats posed by biodiversity loss and the climate crisis (UNDP, 2021, UNEP, 2021, IPCC, 2014a, b). Scientists contend that the scale of the threats to the biosphere and its life forms, including humanity, is so grand that it is difficult to grasp for even well-informed experts (Western, 2021). Climate change impacts conflate with existing inequalities and are currently threatening significant development gains over the past decades, result-

A. Simmons, *Climate Change Adaptation Framework and Youth Entrepreneurship in West Africa*, Sustainable Development Goals Series, https://doi.org/10.1007/978-3-030-85754-7_1

ing in the reversal of development goals. While global leaders focus on the COVID-19 pandemic, the list of global problems is stacking up, including the climate crisis, the collapse of biodiversity and ecosystems and the pollution of the ocean. Reports warn that climate change-induced mass migration, more pandemics and increases in conflicts over resources will be inevitable during the next decade if the current situation continues unchecked (UNDP, 2021; UNEP, 2021).

Humanity is causing rapid loss of biodiversity and the earth's ability to support complex life processes and ecosystems (Bradshaw et al., 2021; Western, 2021). During the past decade, the world failed to accomplish any of the Aichi biodiversity targets (UNEP, 2021; Ripple et al., 2019). It is important to note that this is the second consecutive time that governments worldwide could not meet the 10 years of biodiversity goals (UNDP, 2020; Bradshaw et al., 2021). Even though a wealth of knowledge exists on a range of solutions to address climate change, biodiversity loss and other development calamities, the current scale of their implementation does not match the persistent evidence of the evolution of the loss of biodiversity globally (Cummings et al., 2006). There is a growing body of compelling evidence that continues to evolve on the severity of the threats to human beings' existence on the earth (Rees, 2020).

Researchers blame widespread ignorance for such loss in biodiversity and increased greenhouse gases (GHGs) (IPCC, 2014a, b; UNEP, 2021). The incremental nature of socio-political processes that plan and implement solutions is also blamed for increasing climate change and biodiversity and ecosystems loss. These archaic planning processes are responsible for further delaying actions to address climate change and biodiversity loss. Individual activities, like using fewer plastics or eating less meat, are not enough to address the world's environmental problems today (UN, 2019). It calls for rapid rollouts of initiatives and programmes which produced significant systematic changes to address these fundamental development challenges. If measures are not taken urgently, people will face untold suffering due to the climate crisis (Ripple et al., 2019; Bradshaw et al., 2021; UNEP, 2021).

Due to the severity of the impact of human activities on our planet, we will not achieve the UN Sustainable Development Goals in 2030 unless we radically change our economy and society. Global actions are required to effectively deal with the enormity of the problem, including a) abolishing the idea of continuous economic growth in its current state, b) correctly pricing environmental extremities in the economy, c) stopping fossil fuels use, d) reining in corporate lobbying and e) empowering women and youth to participate in sustainable development (Ripple et al., 2019, UN, 2019, David, 2019).

Another important body of work worth mentioning is the Climate Vote, a collaborated effort between the UNDP and the University of Oxford to produce the People's Climate Vote report in 2021 (UNDP, 2021). The Climate Vote report is the largest survey of public opinion on climate change among over 1.2 million respondents using new and unconventional approaches to polling which spans 50 countries and 56% of the world population. Over half a million under-18 s participated in the People Climate Vote. It is important to note that cohorts represented a key constituent to climate change who have no voice in policy and decision-making because they cannot vote in national elections (Western, 2021).

The Climate Vote provides opportunities for youth to air their voice on the impact of climate change and was undoubtedly a deliberate and conscious effort on the part of the organisers of the survey (Western, 2021). This survey was critical in shaping the research conducted on the 'Impact of Climate Change on Youth in Small Island Communities', which presented the Youth Climate Change Adaptation Framework for use by policymakers, researchers and practitioners when working on youth development and climate change issues (Simmons, 2021).

The survey results state that 64% of respondents believed that climate change is a global emergency despite the ongoing COVID-19 pandemic that impacts all aspects of international life; those surveyed recognised a link between climate change, the COVID-19 pandemic and other development calamities. It is important to note that the COVID-19 pandemic offers an indication of our future in which the stress on the planet reflects the burdens on our future (UNDP, 2020; Brown, 2021; Bradshaw et al., 2021).

Table 1.1 shows the policies that should be put in place to address the climate change challenge. The survey showed a high correction between education backgrounds and the desire to address the climate change challenge. The respondents who attended universities and colleges from both developed countries and developing, low-income countries recognised the current climate change emergency. Although under-18 s were predicted to say that the world is experiencing a climate emergency, researchers were stunned that the other age groups were making similar forecasts (UNDP, 2021).

> 'The results of the survey clearly illustrate that urgent climate action has broad support among the people around the globe, across nationalities, age, gender and education level. But more than that, the pole reveals how people want their policymakers to tackle the crisis. From climate friendly farming to protecting nature and investing in green recovery from COVID-19, the survey brings the voices of the people to the forefront of the climate debate. It signals ways on which countries can move forward with the public support as we work to ether to tackle this enormous challenge'.
> Achim Steiner, UNDP Administrator, 2021 https://www.undp.org/content/undp/en/home/presscenter/pressreleases/2021/Worlds_largest_survey_of_public_opinion_on_climate_change_a_majority_of_people_call_for_wide_ranging_action.html

Electronic games as a vehicle for reaching young people under 18 years old were innovative for collecting the survey data. However, one may argue that many young people from developing countries and states would not have the resources to purchase and/or access electronic games, so their participation would be limited. Nevertheless, Professor Stephen Fisher from the University of Oxford posits that the survey showed that mobilising gaming networks increases the reach and access to audiences and possesses the capacity to engage different groups of people and countries. Besides providing a wide range of data on public opinion regarding climate change, the People's Climate Vote recognises that the climate emergency is more widespread than previously anticipated; most look forward to more robust and wide-ranging policy responses UNDP (2021).

Table 1.1 Climate change policy prioritisation by respondents

Policies	Respondent agreement (%)
Conservation of forest and land	56%
Renewable energy	53%
Adapting climate-friendly farming techniques	52%
Investing in green businesses and jobs	50%

Paris Climate Change Agreement and Future Adaptation Efforts

The Paris Climate Change Agreement is a legally binding international treaty on climate change adopted by 196 parties at COP21 in Paris on 12 December 2015 and entered in force on 4 November 2016. The overall goal is to limit global warming well below 2 °C, preferable to 1.5 °C compared to pre-industrial levels (https://unfccc.int/process-and-meetings/the-paris-agreement/the-paris-agreement)

Adaptation is a crucial pillar of the Paris Climate Change Agreement because it reduces countries and the community's vulnerability to climate change by enhancing their ability to observe the impact. The Climate Change Agreement is a landmark agreement that requires signatories to implement adaptation measures by developing and implementing national plans, climate information systems, early warning systems, protective measures and investments in the green future. Its implementation requires economic and social transformation based on science (https://www.youthapac.org/climate). Parties are responsible for developing their National Development Plans, i.e. National Determined Contributions (NDC). NDC communicates the country's actions to build the resilience to adapt to rising temperature and reduce their country's greenhouse gas emissions to achieve the Paris Agreement's goals (Government of Sierra Leone, 2019a, b; IMF, 2019).

Developed countries are responsible for providing financial assistance to countries more vulnerable to climate change and development calamities. The treaty encouraged developed countries to provide voluntary financial contributions to countries to finance their national plans. However, there were weaknesses identified with the 2015 Paris Climate Change Agreement (Clemencon, 2016). Despite countries' commitment to providing financial resources and technical assistance to enhance poorer countries' institutional capacity and capabilities to implement their obligations to the agreement, these resources are not forthcoming to support climate change adaptation and mitigation initiatives in developing countries. The COVID-19 pandemic has created serious economic and social development stagnations in developed and developing countries alike and possessed the potential to contribute to the shortfall in financial resources required to finance climate change adaptation projects in the future (Bottan et al., 2020; Brown, 2021). This situation is compounded by the absence of human capacities in developing countries to prepare fundable projects to access climate funds.

As temperatures continue to rise and climate change impact increases, countries are making efforts to adapt to the impact of climate change by restructuring their National Development Plans for alignment with National Sustainable Development Goals (Government of Sierra Leone, 2019a, b). However, while countries progress in planning climate change adaptation initiatives, huge gaps remain in financial availability to support climate change adaptation initiatives in developing countries (UNEP, 2021; IISD, 2020). Multi- and bilateral agencies need to increase climate-specific finance to support countries to bring adaptation projects to the stage where they provide real protection against climate change impacts such as droughts, floods and sea level rises. Overall, more effort needs to focus on mobilising private finance to fund adaptation projects to support faster climate action and implementation processes (IISD, 2020).

An essential consideration for moving the climate change adaptation agenda forward this decade is developing natural adaptation solutions that support appropriate local actions that address public and communal changes (UNEP, 2021). The local natural adaptation solutions provide benefits that enhance human well-being and reverse the loss of biodiversity by protecting and supporting the sustainable management of the country's natural resource base and restoring ecosystems. However, this approach has its negatives; it may result in fuelling cultural or human conflicts among population sectors dependent on using natural resources for their survival.

Although the Youth Employment and Entrepreneurship Project (YEEP) was not initially viewed by the Government of Sierra Leone and the African Development Bank (AfDB) as a climate change adaptation initiative, governments, key stakeholders and funding agencies should consider any such initiative by restructuring it, if necessary, to be climate-friendly and build local community resilience to adapt to climate change. Youth development and youth employment are high on the government's development agenda. It is important to note that seven of the eight development clusters of the National Development Plan of Sierra Leone are related to youth development (IMF, 2019; Government of Sierra Leone, 2019a, b). The implementation of the Youth Climate Change Adaptation Framework within the YEEP context provides the consultants with the opportunity to design, develop and implement sustainable development curricula, training manuals and methodologies for students and trainers and support the delivery of skills training at the tertiary level.

Impact of Climate Change in Africa

Africa is considered the most vulnerable continent to climate change, biodiversity loss and other development calamities (Nathaniel & Tetteh, 2017). Evidence from studies stated that climate change has led to the displacement of inhabitants in most countries of the region (Nathaniel & Tetteh, 2017). Africa is the lowest

GHG emission source among all the continents because of the lack of industrial development (Gemeda & Sima, 2015); however, it is deemed the most vulnerable to climate change (Nathaniel & Tetteh, 2017). The most common climate events in Africa are droughts, flooding, desertification and land degradation.

Climate change in Africa is predicted to rise faster than in the rest of the world, exceeding 2 degrees C by mid-century and 4 degrees C by the end of this century (Niang et al. 2014). Climate change is projected to increase temperature and precipitation vulnerability in east Africa. It is causing the nomadic populations to adjust their traditional migration patterns to cope with the impacts of desertification and other climate stressors.

Climate change and environmental degradation are causing displacement and migration, and it is predicted that this will further escalate in the future across the African continent (Godfrey & Tunhuma, 2020). Over 86 million people in sub-Saharan Africa are deemed internal climate migrants. In West Africa, fisherfolks and their families migrate from coastal villages to cities because of the depletion of fishing resources and other supported ecosystems due to ocean acidification and plastic.

These displacements can either be temporary in flooding and landslides or permanently due to loss of arable land caused by desertification and poor agricultural practices and coastal erosion from sea level rise and storms. People are constantly migrating to escape the severe impact of climate change on their well-being and find new sources of sustainable livelihoods. It is also important to note that the impacts of climate change will vary between regions, from sector to sector and even within sectors. Therefore, the knowledge of specific vulnerabilities and risks associated with climate change is crucial for the planning and implementation of climate change adaptation actions at all levels along the development process.

Impact of Climate Change in Sierra Leone, West Africa

Climate change is severely impacting the natural resource base of developing countries in Africa, including Sierra Leone, where the natural resources are limited and fragile. It is attributed both directly and indirectly to human activity, which alters the global atmosphere composition, compounded upon the natural climate variability observed over a comparable period (UNFPCCC, 2012 in IPCC, 2014a, b). Least developing countries (LDC) such as Sierra Leone are severely affected by climate change due to their dependency on agriculture and mining and their insufficient financial, technical, and institutional capacity to adapt (Rose, 2015).

Unregulated and unchecked resource usage has been a cause for concern in Sierra Leone. Mining and agricultural production have accounted for the most dominant economic activities for decades. For example, agriculture generates more than 50% of the GDP in some African countries and employs approximately 60% of the population (Hummel, 2015). However, the use of agrochemicals for commercial farm cultivation has created severe environmental problems. Climate change has compounded these effects through widespread deforestation, wildlife destruction, flash flooding, increased soil erosion, sedimentation and coastal and marine resource damage, especially mangrove, seagrass and reef systems.

Urban centres such as Freetown, Bo and Makeni depend on developing infrastructures such as roads, bridges, power stations and water facilities to maintain the current population and future expansion; however, climatic conditions such as landslides, floods and heavy rains ravaged the city's infrastructure (Nathaniel & Tetteh, 2017). Urbanisation has also resulted in uncontrolled and unregulated construction of squatter settlements on marginalised lands in cities and other urban centres across Sierra Leone, resulting in landslides and deaths. The literature provided an understanding of the possible economic-wide consequences of extreme climate events that affect roads and other infrastructure in Africa, mainly in food security and sustainability (Cervigni et al., 2016; Nathaniel & Tetteh, 2017).

With the declining fortunes of agriculture and mining, the government's policy now focuses on development through diversification into tourism and other economic sectors. Historically, uncontrolled construction development activities have destroyed critical habitats and resulted in beach sand depletion, disruptions to coastal habitat function due to construction close to high-tide borders, and the poisoning of ecological populations and ecosystems due to solid and liquid waste mismanagement generated from rapid urbanisation and squatting.

The fishing sector in Sierra Leone is unregulated; fish exports to nearby countries have increased demand for various fish species and are causing strains on over-fished nearshore fisheries. The fishing sector is further affected by the indiscriminate dumping of garbage, specifically plastic, severely impacting the marine environment; large deposits of plastic and rubbish can be seen on beaches, along gullies and streams, and floating in the sea. Government agencies are taking initiatives to address some of these critical environmental challenges; however, these problems persist and intensify due to weak administrative structures and limited capacity resulting from inadequate human and financial resources to implement policies and programmes.

Innovations to Enhance Human and Institutional Capacity and Efficiency

Building on the body of work on top-down and bottom-up approaches, the human capital framework and other approaches relating to climate change adaptation in volume one (Simmons, 2021), this book explores the relevance and effectiveness of applying new approaches, such as a) the information technologies and information system (IT/IS), b) business process reengineering and c) cultural webs as strategic processes to enhance the capacity and efficiency of youth development institutions during the implementation of the YEEP in Sierra Leone, West Africa.

Information Technologies and Information System (IT/IS)

Numerous institutions, including the National Youth Commission (NAYCOM) and the Ministry of Youth Affairs in Sierra Leone, West Africa, used ICT in the modern workplace. However, these institutions are not equipped with the knowledge and skills to use ICT for strategic decision-making. The IT/IS assesses whether IT is used for making strategic decisions within the organisation.

As technology evolves and expands, so too does the emphasis on its use in organisational changes. In this modern age of technological advancement, organisations acquire the most sophisticated technologies to give them a competitive advantage. The leading-edge focus in businesses today for technology application is now in decision support and competitive advantage. Leading organisations use information technology to support their work at the operational level and, more importantly, incorporate IT at the strategic level. Strategic information system (SIS) is defined as systems designed to bring a competitive advantage or have a competitive edge (Hickman, 2002; Ward & Elvin, 1999).

Therefore, to implement successful change within NAYCOM, management needs to have a clear understanding of the culture of that organisation. Organisations can be captured by their culture, which is the product of history and/or by institutional forces that exist across organisations (Johnson et al., 2017, Johnson & Scholes, 1988, 2002; Porkka, 2016). The consortium utilised the cultural web and BPR to understand the organisation's dynamics and enhance the overall operational processes of the NAYOM to implement the YEEP in Sierra Leone.

Management and Implementation of the Youth Climate Change Adaption Framework

The United Nations defines young people as those between 15 and 24 years old, whereas youth are categorised as those under 29 years of age by the Commonwealth Secretariat. By the UN's definition, young people are estimated at one-sixth of the world's global population, 1.2 billion young people (UNEP, 2021). Youth's engagement in understanding the state of the environment and the importance of protecting it is essential to attain the Sustainable Development Goals promoted by the United Nations. It is necessary to note that young people's role is vitally important in ensuring sustainable development goals from a demographic imperative. Mechanisms must be put in place to ensure that they are given a voice and actively participate in such development (UNEP, 2021; Simmons, 2021).

The development and implementation of a new conceptual framework on youth climate change adaptation provide unique opportunities and appropriate approaches to building youth capacity in small island developing states (SIDS) to adapt to climate change impacts in the future. The framework is divided into four steps: (1) climate change stressor identification, (2) socioeconomic impact assessment, (3) skills and knowledge analysis of youth to enhance human capital adaptation, and (4) youth development policy, strategy, and action adjustments to strengthen capacity and resilience. The framework provides policymakers and researchers with an approach to prepare youth with the necessary knowledge and skills to build their resilience to adapt to climate change and create sustainable livelihoods in small island developing communities. It brings youth development and climate change adaptation to the forefront of the broader global sustainable development agenda (Simmons, 2021).

The framework makes an academic contribution to the global body of knowledge on youth development, sustainable development and climate change adaptation (CCA) within the SIDS development context. This framework could be applied by policymakers and senior officials in SIDS to develop policies and programmes to enable youth to adapt to the impact of climate change. It is also appropriate for conducting further research on vulnerabilities and resilience in disaster preparedness, economic development, tourism, agriculture, fisheries and forestry development. The framework was adapted and piloted in Sierra Leone, an LDC in West Africa. This book highlights the development challenges and accomplishments of implementing the Youth Climate Change Adaptation Framework on YEEP.

Youth Development and Implementation of the Framework in Sierra Leone

Youth in Sierra Leone are defined as young people between 15 and 35 years. Approximately 39.4% of the country's population is youth, of which about 51% are young women (Government of Sierra Leone, 2015). The country has one of the highest youth unemployment rates in West Africa, which stands at 70% (Government of Sierra Leone, 2019a, b). To combat youth unemployment, the government has implemented a series of youth development projects, including youth in agriculture, youth car wash project, youth fisheries project and many others. These projects, however, are encountering financial, administrative and capacity development challenges.

Despite these development challenges, the government's National Youth Commission (NAYCOM) has successfully implemented the YEEP pilot phase with the support of the AfDB, coordinated by the LA International Cooperation-led consortium. The YEEP intends to bolster Sierra Leone's youth's employability through youth skills for employment and entrepreneurship development training. This was accomplished by adapting a Youth Climate Change Adaptation Framework developed initially for SIDS due to similarities in a) the vulnerabilities to the impact of climate change and other development calamities, b) their distance from major economic centres in developed countries, c) high incidence of poverty and other social development challenges, d) low levels of human capacity and technologies and e) youth development challenges affecting both categories of countries. An implementation plan was developed by the LA International-led consortium in September 2018 after implementing a Stakeholders Consultative Conference and submitted to and approved by NAYCOM to pilot the Youth Climate Change Adaptation Framework in Sierra Leone.

The Youth Climate Change Adaptation Framework was restructured and tailored to meet the Youth and Government of Sierra Leone's development needs. The new framework integrates education and vocational institutions with job market demand and supply. It envisaged that this approach would strengthen professional levels in youth population segments and eventually trigger sustained development in the economic sectors. The YEEP has three main components: a) Skills Development for Employment (SDE), b) Career Guidance and Job Readiness Skills (CGJS) and c) Graduate Entrepreneurship Programmes (GEP). The Youth Climate Change Adaptation Framework was relevant to implementing YEEP in Sierra Leone by acting as a model to guide the project implementation process as a management and monitoring and evaluation (M&E) apparatus. The framework was effective in informing the strategic approaches and methodologies developed and implemented during the project. Due to the YEEP pilot project's success, it is recommended that the framework be adopted in the future upscaling of YEEP and supporting other youth climate change adaptation and entrepreneurship schemes around West Africa (Simmons, 2021).

YEEP's overall aim was to reinforce Sierra Leonean youth's employability and enhance their entrepreneurship skills by developing a framework that integrates education and vocational institutions with job market demand and supply.

The intervention has triggered sustained economic development in the growth sectors by establishing youth-led small-scale enterprises. This project was linked to the National Medium-Term Strategy (The New Direction) of the Government of Sierra Leone. Some of the medium-term development strategy's core strategies are youth entrepreneurship, tourism development, agriculture development, mining and human capital development. The government viewed entrepreneurship as a vehicle to help youth develop their communities and the nation by accessing and developing new forms of employment opportunities that reduce poverty in their communities. The initiative is linked to the human capital development programmes administered under the government's flagship education development programme. Initially designed to accommodate 2700 young people, over 6500 youth participated in the employment and entrepreneurship programme implemented by 12 tertiary and TVET institutions during the YEEP pilot phase. After graduating from these skills for employment development and entrepreneurship programmes, youth were prepared to find or create their own employment opportunities (LA International Corporation, 2020).

The Outcomes of the YEEP Pilot Project in Sierra Leone

Despite the outbreak of the COVID-19 pandemic, YEEP activities were implemented in line with the anticipated objectives and achieved the following outcomes: development and approval of curricula by the National Council for Technical Vocational Education Award Agency of Sierra Leone (NCTVA) in Sustainable Construction (Masonry, Carpentry & Aluminium Fabrication), Sustainable Hospitalities (Hotel and Tourism Management), Sustainable Agriculture (Aquaculture and Fisheries Management) and Career Guidance and Job Readiness Skills and Entrepreneurship Development (Graduate Entrepreneurship Programme). The NCTVA is the government agency responsible for the approval of TVET-related development and approval of curriculum.

Implementation of training of trainers for counsellors, instructors/lecturers and administrators from 11 training TVET and tertiary institutions in a) Career Guidance and Job Readiness Skills, b) Skills for Development and Employment and c) the Graduate Entrepreneurship Programme to enhance the delivery of training to youth by partner institutions. The project trained 138 trainers and administrators of universities, colleges, TVET institutions and NGOs in new skills and methodologies to enhance the delivery of instructions to youths. Initially, only 20 officials were earmarked to access this training.

The consortium and government provided tools and financial and technical incentives to partner institutions and support them during the project's implementation.

Training of 6130 youth on career guidance and job readiness skills was conducted. Initially, the project was to train 2400 youths.

Three hundred and twenty-five (325) youths successfully pursued and graduated from TVET courses in Sustainable Construction (Masonry, Carpentry & Aluminium Fabrication), Sustainable Hospitality (Hotel and Tourism Management) and Sustainable Agriculture (Aquaculture and Fisheries Management). Initially, the project was to train 150 youth from 3 partner institutions in the 3 priority skills areas.

Despite poor Internet connectivity, the absence of appropriate technologies and having no experience in using e-learning, during the COVID-19 pandemic, the consortium worked with partner institutions to successfully develop an e-learning mechanism to deliver youth training and technical support to over 6500 young people in urban and rural areas in Sierra Leone. The project was designed to provide training for 100 youths in supervisory management skills; however the consortium trained 104 youths from the 3 tertiary and TVET institutions located in urban and rural centres.

The development of an online platform for graduate entrepreneurs, mentors, business partners and business coaches was successfully implemented. The online platform provides opportunities to enhance partnership among key stakeholders to share information and develop a funding proposal for mobilising financial resources to support youth entrepreneurs across Sierra Leone.

Innovative change management tools that assess youth development institutions' weaknesses and support devising strategies to enhance youth development and organisations' institutional capacity to deliver programs were implemented successfully.

Purpose of the Book

This book is the second in the series exploring the knowledge deficit of known and necessary practices to enhance local resilience to adapt to climate change. The book evolved out of the development consultants' processes to implement the Youth Employment and Entrepreneurship Project (YEEP) utilising the Youth Climate Change Adaptation Framework to implement its youth employment and entrepreneurship initiative in Sierra Leone. The YEEP was funded by Sierra Leone's government through the National Youth Commission by the African Development Bank (AfDB).

One of the significant challenges working in climate change adaptation is the absence of local, sustainable actions with scaling up to regional and international levels. The book contained the outcomes of the challenges, processes, strategies and strategic actions which result in the successful implementation of the Youth Climate Change Adaptation Framework in Africa. The framework was initially developed for SIDS (Simmons, 2021).

While the list of global development problems is growing, including climate crisis, the collapse of biodiversity and ecosystems and ocean acidification, international leaders are battling the COVID-19 pandemic. The COVID-19 pandemic offers us a glimpse into our future if we continue to move forward as business as usual (UNDP, 2021; UNEP, 2021). Due to the interconnectedness of the planet, it took the COVID-19 virus a short time to spread into a global pandemic infecting and killing millions of people. The pandemic is exploiting and exposing the inequalities and the weaknesses in our global social systems resulting in the reversal of the development gains over the past five decades. The recovery from the COVID-19 crisis needs to be more strategic, environmentally sustainable and inclusive. In our effort to reconstruct and rebuild our societies, concerted efforts must be made to ensure that those primitive economic development pathways, practices and behaviours are avoided at all costs. In our reconstruction of global sustainable pathways, the focus should be on the development of policies and actions to strengthen the overall well-being of all humans and species, the reduction of GHG emissions, development of nature-based ecosystem solutions, circular economic practices and behavioural change (UNDP, 2021; UNEP, 2021, Sustainable Development Solutions Network – Youth (2020)).

References

Bottan, N., Hoffmann, B., & Vera-Cossio, D. (2020). The unequal impact of the coronavirus pandemic: Evidence from seventeen developing countries. *PLoS One, 15*(10), e0239797. https://doi.org/10.1371/journal.pone.0239797

Bradshaw, Corey, J. A., Ehrlich, P. R., Beattie, A., Ceballos, G., Crist, E., Diamond, J., Dirzo, R., Ehlich, A. H., Rodolfo, M. E., Pyke, G., Raven, P. H., Ripple, W. J., Saltre, F., Turnbull, C., Wackernagel, M., & Blumstein, D. T. (2021, January 31). Underestimating the challenges of avoiding a ghastly future. *Frontiers in Conservation Science, 1*, 1–10. https://doi.org/10.3389/fcosc.2020.615419

Brown, S. (2021). The impact of COVID-19 on development assistance. *International Journal: Canada's Journal of Global Policy Analysis; Canada International Council*; First Published January 9, 201220211 Release Article.

Cervigni, R., Losos, A. M., Chinowsky, P., & Neumann, J. L. (2016). *Enhancing the climate resilience of Africa's infrastructure: The roads and bridges sector* (No. 110137). The World Bank. Available at: http://documents.worldbank.org/curated/en/270671478809724744/Enhancing-theclimate-resilience-of-Africa-s-Infrastructure-the-roads-and-bridges-sector

Clemencon, R. (2016). The two sides of the Paris climate agreement: Dismal failure or historic breakthrough? *Journal of Environment and Development, SAGE, 25*(1), 3–34.

Cummings, G. S., Cummings, D. H. M., & Redman, C. H. (2006). Scale mismatch in social-ecological systems: Causes, consequences and solutions. *Ecology and Society, 11*, 14.

David, C. (2019). Climate crisis: 11,000 Scientists warns of untold suffering. The statement sets out vital signs as indications of the climate change emergency 5th Novembers, 2019, 15:00 hrs. *The Guardian Newspaper*, UK Edition.

Gemeda, D. O., & Sima, A. D. (2015). The impacts of climate change on the African continent and the way forward. *Journal of Ecology and the Natural Environment, 7*(10), 256–262.

Godfrey, S., & Tunhuma, F. A. (2020). *The climate crisis: Climate change impacts, trends and vulnerabilities of children in Sub Sahara Africa*, United Nations Children's Fund Eastern and Southern Africa Regional Office, Nairobi, September 2020.

Government of Sierra Leone. (2015). *Population and housing census*. Government of Sierra Leone.

Government of Sierra Leone. (2019a). *Youth skills development for entrepreneurship and employment*. National Youth Commission (NAYCOM).

Government of Sierra Leone. (2019b). *The government of Sierra Leone's new medium-term National Development Plan (MTNDP) 2019–2023*. International Monetary Fund.

Hickman, S. (2002). *Handout on strategic information system planning (SISP) – Information technology and business*. Masters in Business Administration, University of Greenwich.

Hummel, D. (2015). Climate change, land degradation and migration in Mali and Senegal – Some policy implications. *Migration and Development, 5*(2), 211–233.

IMF. (2019). *Sierra Leone: Economic development documents-National Development Plan, 2019–23*. International Monetary Fund. African Department, Volume/Issue 2019/218; Published Date: July 2019.

International Institute for Sustainable Development (IISD). (2020). *Filling the gap: A review of multilateral development banks' efforts to scale up financing for climate adaptation REPORT*. Published by the International Institute for Sustainable Development.

IPCC. (2014a). Climate change (2014) synthesis report. In *Contribution of working groups I, II and III to the fifth assessment report of the intergovernmental panel on climate change* (p. 151). IPCC.

IPCC. (2014b). Summary for policymakers. In C. B. Barros, V. R. Dokken, D. Mach, K. J. Mastrandrea, M. D. Bilir, T. E. Chatterjee, M. Ebi, & K. L. Estrada (Eds.), *Climate change impacts, adaptation and vulnerability* (Contributions of the working group II to the fifth assessment report, field). IPCC.

Johnson, G., & Scholes, K. (1988). *Exploring corporate strategy*. Prentice Hall.

Johnson, G., & Scholes, K. (2002). *Exploring corporate strategy* (6th ed.). Prentice Hall.

Johnson, G., Scholes, K., & Whittington, R. (2017). *Exploring corporate strategy: Text & Cases* (11th ed.). Pearson Education Limited.

LA International Corporation. (2020). *Final report of youth employment and entrepreneurship project*. National Youth Commission; Government of Sierra Leone; LA International Corp LED Consortium.

Nathaniel, O., & Tetteh, A. J. (2017, December). The impact of climate change in Africa: A review of literature. *Journal of International Academic Research for Multidisciplinary Impact Factors, 5*(4483). ISSN: 2320-5083.

Porkka, P. (2016). Functional model for organisational and safety culture. *Chemical Engineering, 48*(1), 907–912.

Rees, W. E. (2020). Ecological economic for Humanity's plagues phase. *Ecological Economics, 169*, 106519.

Ripple, W. J., Wolf, C., Newsome, T. M., Barnard, P., & Moomaw, W. R. (2019). World Scientists' warning of a climate emergency. *Bioscience, 70*(1), 8–12.

Rose, R. M. (2015). The impact of climate change on human security in the Sahel region of Africa. *Donnish Journal of African Studies and Development, 1*(2), 9–14.

Simmons, A. (2021). *Impact of climate change on youth in Small Island communities* (Sustainable development goals series, climate change) (Vol. 1). Springer Nature.

Sustainable Development Solutions Network – Youth. (2020). *Youth solution report – 4th edition, sustainable development solutions network – Youth.* UNDP. https://www.undp.org/content/undp/en/home/librarypage/democratic-governance/youth-solutions-report-2020.html

UN. (2019). *UN report: Nature's dangerous decline unprecedented: Special extinction rates accelerating.* Published under UN Sustainable Development Goals, Paris 6 May 2019.

UNDP. (2020). *Human development report: The next frontier, human development and the Anthropocene.* UNDP. Available at: http://report.hdr.undp.org/part-1.html

UNDP. (2021). *World's largest survey of public opinion on climate change: A majority of people call for wide-ranging action.* United Nations Development Programmes. Available at: https://www.undp.org/content/undp/en/home/presscenter/pressreleases/2021/Worlds_largest_survey_of_public_opinion_on_climate_change_a_majority_of_pcoplc_call_for_wide_ranging_action.html

UNEP. (2021). *GEO-6 for youth.* UN Environment Programme.

Ward, J., & Elvin, R. (1999). *A new framework for managing IT- enabled business change* (pp. 197–221). Cranfield, School of Management.

Western, P. (2021). *Scientist warns of ghostly future of mass extension and climate disruptions.* The Guardian Newspaper: The Guardian, 13th January 2021.

2 An Assessment of the Literature on Youth Climate Change Adaptation and Development Calamities and Their Application to Enhancing Sustainable Livelihoods Through Entrepreneurship in Africa

Abstract

This chapter presents current research in youth development related to climate change adaptation globally, specifically highlighting the impact of climate change on developing countries. It gives an updated review of literature that is found in the book "Impact of Climate Change on Youth in Small Island Communities", by identifying the gaps in the literature and justifying the need for the implementation of the study in Sierra Leone, a least developing country (LDC) in West Africa. It identifies the problems associated with climate change which led to the creation of the new framework in SIDS and sets the necessity for testing the Climate Change Framework in West Africa.

The chapter positions the book within the context of human capital theory and highlights the current research findings from the IPCC on climate change and biodiversity loss and its impact on youth. The literature review further explores the context of human capital theory and its links to climate change adaptation. This is done within the framework of youth in a west African context, Education and Skills training for Climate Change Adaptation (CCA), and e-learning and modern technologies as an instrument for responding to development calamities such as the Ebola epidemic and COVID19 pandemic in LDCs.

The chapter further builds on the relevance of the top-down and bottom-up approaches for the implementation of youth skills development training for employment and sustainable livelihoods (Simmons A, Impact of climate change on youth in small island communities. Sustainable Development goals series, climate change, vol 1. Springer, Cham, 2021). The Chapter reviews the TVET programmes in Africa and explores a range of methodologies for training to enhance the delivery of training in Technical Vocational Education and Training (TVET) in West Africa. It also explores the use of Business Process Reengineering and Information Technologies Systems for enhancing the efficiency and capacity of youth development institutions.

Keywords

Climate change · Sustainable development · Youth development · Entrepreneurship · Employment · Climate change adaptation · Africa · Sierra Leone · biodiversity · Technical vocational education training · Conceptual framework · Top-down approach · Bottom-up approach

This chapter identifies the gaps in the literature and justifies the need to implement the study in

A. Simmons, *Climate Change Adaptation Framework and Youth Entrepreneurship in West Africa*, Sustainable Development Goals Series, https://doi.org/10.1007/978-3-030-85754-7_2

Sierra Leone, a least developing country (LDC) in West Africa. It identifies the problems associated with climate change which led to creating the new framework in SIDS and sets the necessity for testing the climate change framework in West Africa. The chapter positions the book within the context of human capital theory and highlights the current research findings from the IPCC on climate change and its impact on SIDS youth. The literature review further explores the context of human capital theory and its links to climate change adaptation, specifically within climate change adaptation education for sustainable development. This is done within the framework of youth development in a West African context, education and skills training for CCA and e-learning and modern technologies as an instrument for responding to development calamities such as the COVID-19 pandemic and Ebola epidemic in LDCs. It explores the relevance of top-down and bottom-up approaches towards the implementation of CCA training for youth employment and sustainable livelihoods within the context of Technical and Vocational Education and Training (TVET) in West Africa.

Climate Change Scenarios and Effects on Youth

Climate change can produce significant impacts across exceptionally large temporal and spatial scales (Marshall & Johnson, 2007; Hernandez-Delgado, 2015; Nurse et al., 2014, IPCC, 2014a, b) with significant adverse consequences on local economies and livelihoods, particularly on the youth sector in small island developing communities/states (Mills, 2005; Ripple et al., 2019).

The IPCC (2014a, b; UNEP, 2021) declares that warming of the climate systems is unequivocal and that most of the observed increase is highly likely due to the increase in anthropogenic greenhouse gas concentration and that the growing accumulation of greenhouse gases (GHGs) in the atmosphere resulting from human activities is exceeding the historical levels that keep the earth liveable (UNEP, 2013). The IPCC developed a series of climate change scenarios which show the change in global mean surface temperature, atmospheric CO_2 concentration and global average SLR under three climate change pathways (Arnell et al., 2016). Under the business as usual (BAU) scenario, continuously rising emissions could drive observation and concentrations of greenhouse gas in the atmosphere from present-day levels of 390.5 ppm CO_2 to 685 ppm with predicted warming of 1 °C–above 3 °C (IPCC, 2014a, b). This will have a severe impact on youth in LDSCs such as Sierra Leone. The range in future temperature projection would impact severely on youth livelihoods, health and overall well-being (Pereznieto et al., 2011).

Climate Change and Possible Manifestations

The global mean temperature is projected to rise over the twenty-first century if greenhouse gas emissions continue unabated (IPCC, 2001, 2014a, b; Ripple et al., 2019). The Special Report on Emissions Scenarios (SRES) A1B medium-emission scenario suggested a 1.8 °C to 2.3 °C medium annual increase in surface temperature in the Caribbean Sea by 2100 when compared to 1980–1999 baseline. A1 scenarios explore the impacts of CC regarding a more integrated world, with A1B evaluating all energy sources. The research shows that this will result in an annual decrease in precipitation of around 12% in the Caribbean region. Comparatively, the projection for the RCP4.5 scenario suggested about 1.2 °C to 2.3 °C increase in surface temperature by 2100 compared to the 1980–1999 baseline. It shows a decrease in precipitation of about 5 to 6% in the Caribbean and Mediterranean regions. Both scenarios indicate that there will be future potential problems for agriculture production and water harvesting in the Caribbean region and in some parts of Africa (IPCC, 2014a, b; Nurse et al., 2014). This will have serious consequences for employment and the overall livelihoods of young people in that a large percentage of them depend on agriculture and tourism for their livelihoods, and there is compelling evidence on the negative impact of climate change on tourism and other sectors.

It is important to note that possible increases in temperature at about 2 °C temperature by 2100 would have far-reaching consequences for ecosystem such as coral reefs, mangrove swamps and other marine and terrestrial ecosystems. Researchers conclude that coral reef is important to the economy and overall livelihoods of SIDS and coastal areas of Sierra Leone, West Africa, specifically fishing and tourism (Bueno et al., 2008; IPCC, 2014a, b, c; Hernandez-Delgado, 2015; Nurse et al., 2014) and that a large percentage of youth are employed in these sectors. Nurse et al. (2014) predicted that in the future the current accumulation of greenhouse gases (GG) in the atmosphere could cause more than half of the world coral reefs to experience frequent thermal stress by the year 2080. Such a rise in temperatures in the Caribbean and on coastal areas of Sierra Leone, West Africa, will be devastating to the tourism industry and health system. Therefore, as temperatures are projected to increase in the future (IPCC, 2014a, b), tourists may choose not to travel to tropical destinations (Benjamin, 2009).

Scientist argued that climate change can affect human health negatively since it may modify the transmission of diseases such as cholera, malaria and meningitis (Nathaniel & Tetteh, 2017). Climate change possesses the capacity to aggravate the incidence and intensity of impending disease outbreak. There is evidence that increased temperature will allow for the increase in vector diseases and acute respiratory infections (Nurse et al., 2014; Hernandez-Delgado, 2015). Poschen (2015) supported the view that there will be an increase in malaria, dengue fever and other diseases because of an increase in climate change. Besides putting more strain on the already overstretched health system in the Caribbean and most countries in Africa, increase in diseases will exacerbate the economic strains in these regions due to a reduction in tourism revenue. Tourist would not come to these countries if there were an increase in vector diseases. This will also put further strains on the human capital, in that large numbers of youth would not be able to participate in productive economic activities due to sickness and those who are available to work would not find any employment in the sector because of the reduction in tourist in these countries (Benjamin, 2009; Hernandez-Delgado, 2015).

A further rise in temperature will negatively affect the marine industry, thus affecting livelihoods such as tourism and fisheries. As discussed earlier, the temperature rises of 1–2 °C can lead to coral bleaching and the destruction of valuable ecosystems (Benjamin, 2009; Nurse et al., 2014; Hernandez-Delgado, 2015). Therefore, loss of these reefs would have detrimental effects on tourism and fisheries in the African region, thus increasing the already high youth unemployment problem in West Africa.

Flooding and Sea Level Rise

There is evidence that difference in climate impacts between 1.5 °C and 2 °C is most pronounced for vulnerable regions and groups, specifically for youth with limited adaptive capacity (Schleussner et al., 2016, IPCC 2014a, b). The impact of SLR would not be uniform in coastal nations in Africa although some countries are projected to experience severe impact from 1 m SLR. Such differences in the vulnerability of these countries can be explained in the geophysical composition of the landscape and their different coastal makeup, specifically those countries with large coastal plains with low lying coastal areas close to sea level such as Sierra Leone, Liberia and Ghana in West Africa. These countries are highly vulnerable to SLR.

Saltwater intrusion into the thin lenses would contaminate the groundwater on coastal areas in Africa. This is a major challenge for these developing countries and their population, which is predominantly youth. Flooding of coastlines and coastal plains is the main effect of SLR, storm surges and tsunamis.

Natural Resources and the Environment

Under all climate pathways, it is estimated that plant and animal habitat would become unsuitable for more than 75% of species due to climate change throughout the twenty-first century. Over 50% of plant species are projected to lose more than 50% of their climatically suitable habitat at 3 °C and over the pathway. Other species would suffer from climate change including a higher proportion of amphibians, a low proportion of birds and mammals and reptiles. Therefore, the impact of climate change on the natural resources base of SIDs and LDCs will be devastating (Simpson et al., 2010; Nurse et al., 2014; Bradshaw et al., 2021).

Climate change has also caused many terrestrial, freshwater and marine species to shift their geographic range, seasonal activities, migration patterns and species interactions in response to the ongoing impact of this phenomenon (IPCC, 2014a, b, c). This severely impacts people living in semi-arid lands, in low-lying coastal areas, in water-limited or flood-prone areas or on small islands (IPCC, 2001; Olmos, 2001). Although data are absent on the impact on youth, the researcher can conclude that any change in the natural resources base of the LDC in Africa would negatively affect the youth.

Climate change is affecting ecosystems and the essential services they are providing to protect flooding and sea level rise. Therefore, climate change tends to devastate biodiversity and wetland regions, leading to soil loss and trees (UNDP, 2021; UNEP, 2021; Bradshaw et al., 2021). Poor and often vulnerable and marginalised communities who are mainly dependent on the services provided by these ecosystems are primarily affected by the negative impact of climate change on these ecosystems (Nathaniel & Tetteh, 2017).

Hurricane and Storms

Climate change-related risks from extreme climatic events such as storms and hurricanes are already moderate (high confidence) and high with 1 °C additional warming (medium confidence) (IPCC, 2014a, b). Consequently, the risk associated with such extreme climatic events would increase further at a higher temperature. People, specifically youths, are particularly vulnerable to the impact of climate change. Storms are projected to increase in frequency and intensity in SIDS and across the continents of Africa in the future (Gomez, 2013; Nurse et al., 2014; The Economist, 2017; Fieser, 2017 Ripple et al., 2019). Climate change and harsh weather conditions such as high temperature and severe rainfall events are factors initiating malaria epidemics in Kenya, Uganda, Ethiopia, Tanzania, Rwanda and Madagascar (Nathaniel & Tetteh, 2017). Climate change has also led to the displacement of the poor and vulnerable segments of the population in some African countries. There is increasing evidence that climate change is affecting forest and forest ecosystems in Africa, especially the livelihoods of forest-dependent communities and the natural economic activities they depend on to eke out their livelihoods (Nathaniel & Tetteh, 2017). The drawing of feasible conclusions on the climatic extremes is a constant worry for researchers due to the lack of a viable body of data. This prevents researchers from determining which storms were caused by natural viability from those that were part of a long-term climatic trend. Another challenge facing researchers is the suitability and applicability of climatic models.

Climate models are not yet sophisticated enough to predict tropical storms which limit the ability of researchers and policymakers to anticipate future changes in the frequency and intensity of hurricanes. These climatic models are primitive regarding the complexities of the atmospheric processes that are involved in the creation of storm systems. Therefore, these models are weak in predicting future climatic events. There is also evidence of increases in stronger storms and other climatic events across the African continent (Nathaniel & Tetteh, 2017). Overall, storms continue to be a costly undertaking for all regions. The cost attributed to storms is projected to increase from 1.9% of GDP in 2025 to 2.1% GDP annually in 2100 on the low-impact scenario, while on the high-impact scenario, it is projected to move from 3.1% GDP in 2025 to 10% GDP in 2100 (Bueno et al., 2008). Storms are destroying housing and adversely affecting the overall well-being of youth and the communities annually in the tropics. Many young people live in poorly constructed housing stocks because they are either unemployed or do not have the financial resources to build substantial houses to withstand the high level of storms.

Youth Development and Employment

Climate change is having a devastating impact on all areas of livelihood, specifically those relating to youth employment (Simmons, 2021; Poschen, 2015; ILO, 2010; UNDP, 2015). A substantial review of the body of literature on youth entrepreneurship and employment in Sierra Leone and across the globe was conducted to gauge the new and emerging models, methodologies and practices in youth employment and sustainable livelihoods. Over 50 documents on the topic were reviewed. Those documents from Sierra Leone included (a) the 2014/15 Status of Youth Report, (b) the blueprint on National Youth Programme 2014–2018, (c) the Youth and Stakeholder Perception Surveys and (d) Research on Selected Key Youth Development Indicators commissioned by NAYCOM with support from the UNDP and the Diagnostic Study of the TVET Sector in Sierra Leone which was commissioned by GIZ and the Government of Sierra Leone (Government of Sierra Leone, 2014, 2015, 2019).

At the international level, relevant literature discusses issues pertinent to entrepreneurship, employment, training and skills acquisition and development in developed and developing areas of the world. Research by ILO and UNDP states that the unemployment problem as currently experienced is further compounded by the impact of climate change and other calamities on the natural and human capital of countries and the productive sectors such as agriculture, tourism, fisheries, mining and others within a county or region of the country (ILO, 2013; Poschen, 2015, UNDP, 2015).

Research has shown that climate change will bring about serious changes to existing enterprises and the economic sectors across the globe, in that some jobs and livelihoods that are currently available in developing countries would not be in existence in the future (ILO, 2013; Poschen, 2015). Jobs that would still exist in the future will change tremendously and require new knowledge and skills to efficiently carry them out. Therefore, building the adaptive capacity of youth to the impact of climate change and other development calamities would be essential to protect enterprises, workplaces and communities from the negative effects of climate change and other calamities (Poschen, 2015; UNDP, 2015; ILO, 2010, 2013; Nurse et al., 2014)) as well as in exploring new areas of economic growth.

One of the major challenges faced by youth in Sierra Leone is the poor quality of education provided by the education institutions responsible for providing training to young people. Carter (2008) identified a mismatch between the skills and knowledge offered by education and training in developing countries and those required by the market/workplace. The mismatch in education is also highlighted in a study by the World Bank on 'Quality Education counts for skills and growth', which states that despite spending nearly 11 years in education, school leavers often struggle to find formal employment (World Bank, 2013). The UNDP (2015) states that the skills mismatch in the knowledge economy is due to the slow pace of technological innovation in developing countries, compounded by the rapid growth in demand for new and higher skills in most developing countries.

The reports also highlighted the role that technological innovation plays in taking up work done by humans and argued that countries can no longer look to manufacturing to absorb their army of unemployed youth since computers and robots now do these works. Overall, the situation can be worsened by climate change in developing countries such as Sierra Leone, specifically on youth and their respective communities.

The educational institutions in most development countries in Africa including Sierra Leone do not equip young people with adequate knowledge and skills to meet the country's current development needs of the region nor prepare them to meet future needs (UNDP, 2015;

Simmons, 2021). The key to fill this gap is to ensure strong interaction between the world of work and education and training. Youth needs to acquire adequate skills and knowledge that would equip them to participate in initiatives relating to the greening of the economy and the use of technology and enhancing their resilience against climate change and other calamities such as the Ebola epidemic and COVID-19 pandemic.

An essential element of the overall education policy is ensuring that all youth can access quality education and training to transition to new jobs or work with new materials, processes and technologies in their existing jobs (Poschen, 2015). It is important to note that effective education and training cannot be achieved by implementing one-of-a-kind or one-off training activity, but a continuous education and training process is essential to achieve the above. It is no longer effective to train youth to meet the current needs and develop skills and knowledge that support entrepreneurship, resilience, innovation in enterprise development and their transition to sustainable practices (Poschen, 2015; ILO, 2013). Government and training providers must ensure access to training programmes that support lifelong skills development and focus on future market needs (ILO, 2010; Poschen, 2015). The Youth Employment and Entrepreneurship Programme (YEEP) implemented in Sierra Leone supports a lifelong skills development model that focuses on the needs of the market.

Some countries in Africa lack the country-specific industrial structure and are not at the appropriate development stage to support the transition of skills and occupational development necessary to create sustainable livelihood required to fill the employment needs of these countries (UNTAD, 2018). As highlighted earlier, there is an existing mismatch in the education system in most developing states between the current education and training provided and the skills required in the job market in the state and/or the region (Carter, 2008, UNEP, 2013), World Bank, 2013). Digital technologies and advanced communication technologies penetrate all areas of life in all developing countries; however, many youths in Sierra Leone do not have training nor access to these new technologies to enable them to live fulfilled lives.

A shift in education and training is urgently required for transit youth to achieve sustainable development goals by providing them with viable skills and knowledge for the future. Training and skills development is understood in broad terms to cover the following: (a) basic education (providing the basis for youth development, laying the groundwork for access sustainable employment/livelihoods); (b) preliminary training offering core work skills, general knowledge and industry networking enabling the shift from education to work; and (c) lifelong learning contributing to the upkeep of individual skill supply (ILO, 2010, UNDP, 2015, Simmons, 2021).

A large percentage of youth from developing countries such as Sierra Leone are unable to acquire appropriate skills (employment and life skills) to find viable decent employment to contribute to their society's development (Carter, 2008; ILO, 2010; UNEP, 2013). These countries are experiencing serious economic challenges resulting from the global financial squeeze, compounded by climate change and the severe impact of COVID-19 pandemic. They are also unable to generate enough resources to enable them to contribute to enhancing sustainable development.

This period is also emerging as a time when flexibility, adaptability and unconventional work become the dominant development practice of the day (UNDP, 2015). It is necessary for governments and key stakeholders to put measures in place to ensure the development and implementation of pathways to alternative skills and knowledge development to prepare youth to access education and training relevant to the changing markets in the countries (ILO, 2010; UNESCO, 2014). Therefore, the Sierra Leone Youth Entrepreneurship and Employment Project (YEEP) is consistent with the literature on employment, training and youth development and should be viewed as a step in the right direction implemented by the Government of Sierra Leone to create decent and sustainable livelihoods for its young sector of its population.

Rationale for Investing in Young People

Youth development is viewed as the period of transition from childhood to adulthood i.e. it is the stage where an individual develops competencies and acquires positive attitudes and personal and job skills to become a productive adult who is well-integrated into the economy and society (World Bank, 2007, UNEP, 2021). Therefore, events that occur during these formative years may lead to behaviours that impede human and social capital investment during this critical period, thus producing individuals who are not well prepared for the challenges of adulthood. This has clear implications for the individual since an ill-prepared adult will have more difficulty finding a job, earn lower wages, and be more likely to face poverty and the attendant social problems (Takewira & Simmons, 2004; Simmons, 2006).

Additionally, an individual whose development is stunted would not be able to positively contribute to the development of the economy and the society at large. Furthermore, lack of skills to contribute positively to the modern economy will result in the following: (i) impede economic growth and exacerbate income inequality and poverty, (ii) a low-skilled work force will not attract foreign investment, (iii) a society with high crime rates will discourage development within export processing zones and tourism, and (iv) an unemployable labour force, high fertility rates, and violence would divert resources away from the productive public investments. Young people who face the challenges of adulthood unprepared are more likely to pass on to their children their negative behaviours, thus perpetuating the vicious cycle. On the other hand, well-developed young people are a public good that create positive externalities, implying a clear role for more involvement of the government (i.e. society at large) and the business sector which would benefit greatly. Consequently, it makes economic, social and political sense for the governments of Africa, in particular Sierra Leone, to invest in youth at the front end rather than waiting to spend huge amount of resources to address the many challenges at the back end such as building more prisons; employing more police and officials to manage the penal system and to fight the drugs cartels, gangs and 'yob' cultures and implement rehabilitation centres for drug users and young offenders; and addressing the spiralling HIV/AIDS problem (Simmons, 2006, 2021).

In emphasising the point pertaining to the necessity of investing in youth, research maintains that youth risks need to be highlighted, and specific programmes should be designed to address the unavoidable risks youth face. Programmes for youth ought to be envisaged as major investment trust that plans for youth as assets and not as liabilities, purposeful investments that would generate guaranteed major benefits and not loss to the economic sphere. Therefore, the development of youth policies should incorporate mechanisms and systems that would minimise the vulnerabilities of youth in the developing countries (Commonwealth, 2007; Simmons, 2006, 2021).

Top-Down and Bottom-Up Approaches to Climate Change Adaptation

Climate change adaptation (CCA) refers to adapting human-driven adjustment in ecological, social or economic systems or policy processes to actual or expected climate stimuli and their effects or impacts (UNFCCC- LEG, 2011 in UNFCCC- LEG 2017). Therefore, various types of adaptation can be distinguished, including anticipatory and reactive adaptation, private and public adaptation and autonomous and planned adaptation (IPCC Fourth Assessment Report (AR$), 2007 in UNFCCC- LEG 2017; IPCC, 2015). Although international efforts to reduce greenhouse gas emissions are increasing, global efforts to limit greenhouse gas emissions are insufficient to avoid the risk of climate change and sea level rise. Climate change adaptation is necessary to help individuals, communities, organisations and natural systems deal with climate change consequences that cannot be avoided. CCA entails taking practical actions to manage risks from climate impacts, protect communities and strengthen the economy's resilience to climate change's damaging effects.

Top-down and bottom-up approaches to CCA were borrowed mainly from business management studies by climate change practitioners and academics. Both strategies have been widely accepted and dominated by empirical studies on the process of operations strategy (OECD, 2013).

Top-Down Approach to Climate Change Adaptation

From a business operational perspective, a top-down process can be described as planned coordination of intention and action to achieve specific outcomes imposed by the central authority such as government (Matland, 1995; OECD, 2013). Top-down CCA is more applicable to government agencies.

Despite all the advancements in technology, there is a high level of uncertainty surrounding climate science. Therefore, due to the inability of policymakers in SIDS to work with the high levels of uncertainty involved in top-down adaptation approaches, much of the research in this field stopped at the impact assessment stage. It is important to note that top-down approaches to climate change adaptation are also mainly used by developed nations because they are less prone to the impacts of and are more resilient to climate change.

Top-down approach has its share of criticism. Some critics claimed that the top-down approach takes statutory language as the starting point and do not consider the significance of previous actions taken to address the issues. Adger et al. (2005) argued that the command-and-control approach has come under scrutiny for its contribution to potential maladaptation exacerbating vulnerabilities and causing negative externalities. The approach is also criticised for considering implementation as an administrative process and usually ignores or eliminates the political aspects of the process. It is also criticised for only taking into consideration the views and decisions of the central decision-makers while neglecting the views of local actors and other stakeholders (Hjem & Hull, 1982; OECD, 2013).

Bottom-Up Approaches to CCA

Bottom-up approaches to CCA are more applicable to projects developed by youth and local communities. Over the past decade, interest was resurgent by international development agencies and practitioners to provide direct support to youth and community-led bottom-up development initiatives to build the capacity of communities to implement community-based climate change adaptation actions (Jaja et al., 2016; McNamara & Buggy, 2016). Bottom-up approaches to CCA are focused mainly on the concept of vulnerability. The approach assumes that if one can address actual vulnerability today, one inevitably reduces future (expected) vulnerability (Burton et al. 2002). Bottom-up approaches consider vulnerability as representative of social and ecological structures that are generated by multiple factors and processes (OECD, 2013).

Bottom-up approaches are more applicable to developing nations including SIDS because much of the structural vulnerability addressed through bottom-up approaches is less of a consideration in developed nations, i.e. the infrastructures needed to adapt to climate change are already in place in developed countries. For adaptation to be successful, it must fundamentally be linked with key development areas such as food security, robust education and health systems and an improvement in infrastructure provision. Therefore, achieving these development goals will help in reducing vulnerability to climate change (Hernandez-Delgado, 2015).

On the issue of relevance to policy development, academics who support the bottom-up approach claimed that the approach emphasises target groups and service deliverers as the main target of development, arguing that policy is being made at the local level and not by centralised government machinery (Matland, 1995). Bottom-up approach develops a process and procedures for implementation of the approach by identifying the network of actors and stakeholders who are involved in the delivery of services at the local area level and asking them about their goals, strategies, activities and contacts (Stochowiak et al., 2016). It then uses the contacts to develop networking techniques to identify the local, regional and national actors involved in the planning, financing and execution of the relevant governmental and non-government programmes.

Some of the strengths of using bottom-up approaches include focusing on the centrally located actors who develop and implement government programmes and policies. The approach is more flexible than the top-down approach in that it does not present prescriptive advice but describe the factors causing difficulties in reaching stated goals (Matland, 1995), i.e. ensuring that strategies are flexible to enable them to adapt to local difficulties and contextual aspects of the respective programme (OECD, 2013). Despite these strengths and benefits, the bottom-up approach is faced with its share of criticisms, including (a) policy control should be implemented by officials whose powers come from their accountability to the sovereign voters through their elected officials, i.e. in a democratic context. The policy should be enacted only by those who have been endowed to exercise power by the voting public, and (b) it overemphasises the level of local autonomy (Matland, 1995; OECD, 2013).

Hybrid Approach: Combining Top-Down and Bottom-up Approaches

There is increasing interest in the literature for combining the micro-level variables of bottom-up and the macro-level variables of the top-down approaches in the development and implementation of climate change adaptation projects to benefit from the advantages and strengths of both approaches by enabling the various levels of stakeholders to interact and influence project outcomes (Matland, 1995; OECD, 2013). Suggett (2011) builds on the work of Matland (1995) which draws on strategies of both approaches to respond to policies that are characterised by high conflict and high ambiguity.

Combining both approaches would highlight their strengths and minimise their weaknesses. Policy implementation is most likely to be successful when all stakeholders at various levels can interact with each other. This provides adequate opportunities for both central policymakers and local stakeholders such as youth and community leaders on the ground to work with each other. This situation is extremely important for the successful implementation of policy and strategic programmes (OECD, 2013).

The hybrid approach is more applicable to the implementation of climate change policies and programmes (OECD, 2013, Stochowiak et al., 2016). These policies are characterised by high levels of ambiguity and conflict and are designed to emphasise value or significant political shift. Such policies tend to have a referential goal and ambiguous plan of action; stakeholders' interest is often tied to policy ambiguity in their group's favour. Therefore, in bottom-up approaches, the strength of the coalition at the community level often contributes to the achievement of outcomes. While the successful outcomes of policy implementation are determined to a large extent on successful outcomes at the community level, the centralised government implementation still plays an important role in policy implementation. Hence centrally located officials can exercise their influence by providing resources and incentives, focusing attention on an issue, establishing strong leadership around a vision for the policy and engaging with networks of local groups and communities (Stochowiak et al., 2016). Stochowiak et al. (2016) argue that the lack of tangible implementation common in high ambiguity/high conflict policy environment can suggest to policymakers and local officials the need to reduce either the level of the conflict or the ambiguity to move the policy to a more successful implementation.

Human Capital as a Vehicle for Climate Change Adaptation and Youth Employment

Climate change and other development calamities such as COVID-19 impact negatively on human capital, specifically on education and health, thereby causing further marginalisation and poverty for the poor. These can destroy livelihoods, thus increasing unemployment and poverty among people who have already been experiencing social and economic development challenges (Qaisrani, 2015). It can also result in increased migration and urbanisation of the economically poor, specifically youth who are already stressed by the impacts of storms, drought, SLR, high temperatures and other climate pressures (UNDP, 2015) (Fig. 2.1.).

Research shows that education is the key to improving resilience to climate change as well as the attainment of global development goals (UNEP, 2013; UNESCO, 2015). Pressures from global warming will make it more challenging for developing countries, especially SIDS, to achieve existing development targets for health and education. However, better quality education would help young people to develop a clear understanding on how to cope with and respond to changes in climatic conditions. Article 6 of the United Nations Framework Convention on Climate Change (UNFCCC) calls on governments to develop and implement programmes on education, training and awareness to build the resilience of people to adapt to the impact of climate change (UN, 2014). This view was supported by Agenda 21 of UNCED which focuses on the reorientation of education towards sustainable development; increasing public awareness; and promoting training (UN, 1992).

Research has shown that as natural hazards and disasters increase, the accumulation of human capital, as measured by secondary school enrolment, would fall. Through education and training, people can be provided with climate-specific know-how and information (knowledge and skills) for improving agriculture, fisheries and tourism performance. The strategy can be

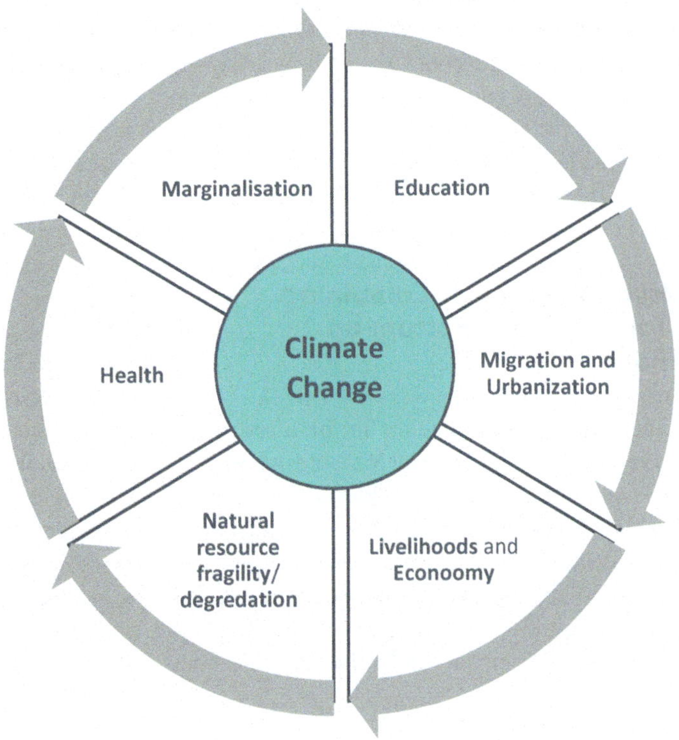

Fig. 2.1 Dynamics of climate change impact on human capital. (Adapted from Qaisrani, 2015)

further enhanced by (a) improving occupational safety and health, (b) the responsible use and management of pesticides and other hazardous chemicals, (c) investments in extension services, (d) provision of support for rural infrastructure and (e) development of a non-farm rural economy (ILO, 2013; IPCC, 2014a, b; Poschen, 2015).

Human capital can also be severely affected by poor health outcomes because of the occurrence of invasive tropical diseases such as malaria, dengue fever and others such as Ebola and COVID-19 (Hernandez-Delgado, 2015; Nathaniel & Tetteh, 2017). These diseases limit the country's ability to develop as well as inhibit the capacity to deal with climate shocks. Therefore, it is imperative that solutions to address the human capital challenge must be built on the local development system and develop in cooperation with the local people such as the youth, women and children (ILO, 2013).

The organisation of youth, farmers, workers and other key stakeholders is an important element to give local communities and other stakeholders a voice in the policymaking process for the overall development and the greening of economies of LDCs and the overall sustainable development of African states. Within this emerging context, emphasis must be placed on the inclusion of women, youth and other marginalised sectors of the population into the local and national development process (ILO, 2013). The development of civil society groups such as cooperatives and community-based organisations would be an important strategy to enable youth to acquire the relevant skills and training to participate effectively in climate change adaptation initiatives at all levels (ILO, 2013; UNEP, 2021; Sustainable Development Solutions Network – Youth, 2020).

The Concept of Quality of Education in the Context of Climate Change Adaptation and Resilience Building

Education is viewed as an essential element of the global response to the impact of climate change (Cordero et al., 2020). As advocated in Chapter 36 of Agenda 21 (UN, 1992), education and awareness-raising make it possible for humanity to indulge in informed decision-making, playing a critical role in increasing adaptation and mitigation capacities of communities, hence empowering women and men to adopt sustainable practices and actions. In many West African communities, exposure to extreme temperature and/or rainfall in early life is associated with fewer years of schooling in later childhood (Paytan et al., 2017). Rising temperatures and fluctuating rain patterns due to climate change may mean more children, and youth, losing out on schooling opportunities without intervention.

Coastal communities in West African regions experiencing a faster rate of coastal warming depend heavily on marine ecosystems and fishery industries to sustain their livelihoods. These communities may need to adapt to a warmer climate and ocean to ensure their food security and future economic survival. It is also important to note that the fisher communities in some of these countries are of low socioeconomic status and will be more vulnerable to the effects of climate change. These communities are vulnerable to the impact of climate change because of the higher percentage of the population who are dependent on fishery industries for their livelihoods and a lower capacity of the population to adapt to climate change and other calamities.

Therefore, the fusion of the increase in rate of temperature rise combined with poverty makes the case for the critical necessity for public education and outreach in these populations as a strategy to build their resilience to adapt to climate change. It is essential to note that efforts to increase climate literacy initiatives are significant to better prepare citizens to adapt to climate change and respond adequately to building their resilience to deal with the economic and environmental challenges confronting their daily lives. Education and training also equip them to exploit the new opportunities that will be created by climate change (Poschen, 2015; ILO, 2010).

The education and outreach initiatives geared to enhance climate literacy must focus on giving humanity the scientific basis to make informed decisions and actions and actions to adapt to the climatic impact and to prepare for future changes.

The Rights-Based Approach to Quality of Education

UNESCO, in a document entitled *Quality Education for All: A Human Rights Issue* (UNESCO, 2007), presented a rights-based approach applied to the development and implementation of education. This approach has implications regarding the understanding of what constitutes quality education for sustainable development. From the point of view of its purposes, it states that education must be *relevant*, i.e. it must enable people to fully exercise their freedom and citizenship. In essence, education can only achieve this goal if it is *pertinent* to the concrete conditions within which people act, i.e. having a universal character by being a right for all, and that it enables people to learn life skills. Therefore, the rights-based approach to education ensures that *equity* is a factor that is inherent to quality.

Moreover, considering that education is a right, public action is of crucial importance. Quality education must be *effective* in achieving the objectives proposed in the field of public action while respecting the resources that citizens devote to the task by being *efficient.*

The tenets of the philosophy of education include the following:

- Education is a fundamental human right, and all citizens should be enabled to achieve a basic required minimum.
- Education should not be limited by age, time and place but should be a lifelong process which goes beyond formal schooling.
- Maturation and social development in the formative years of the life of the individual should be directed by the combined influence of the school, home and society.
- Given the role of education in social life, both universal and compulsory education is necessary at certain periods in the life of every citizen. All forms of education, whatever the content, should be developed to meet the varying learning needs of the population.
- A morally sound concept of the equality of educational opportunity should be central to the thought and practice of education.

Over the past two decades, curricular frameworks have moved from the focus-centred transmission of knowledge approach to those that foster in students the construction of knowledge, the processing and analysis of information and the development of critical thinking (UNESCO, 2007). Good learning takes place in interactive processes and contexts – not in unidirectional transmission processes in which students are able to exchange their hypotheses, ideas and meanings with what appears in books and other materials, including the Internet, and with what their teachers and peers state. From the analysis, it is possible to observe that most curricular principles express the idea of mobilising and applying knowledge – or rather the development of conceptual, procedural and attitudinal content to enable young people to perform satisfactorily in the world.

Quisumbing (2001) views quality of education in terms of quality of the teaching- learning outcomes. Quality in this sense consists mostly of skills and competencies needed for a particular group in society. In this sense, quality of education is related to descriptive and evaluative (quantitative) and to some extent qualitative outcomes.

The 1990 Jomtien Declaration Education for all defines quality education not only as quantifiable outcomes such as pass rates, but it also embraces values and changes in attitudes that are not easily measured but nevertheless observed. This definition of quality of education includes both quantitative and qualitative outcomes, unlike those above.

Over the past decade, the value of education for climate change adaptation policy has been limited by vague definitions and poor communications practices. This is due to the small amount of research in this area of study. The greatest value of education lies in the transformative potential of adaptation learning support to (a) curricular development, (b) the development of pedagogical processes and practice and (c) technological resources that prepare people for complex adaptive decision-making and help them enhance learning.

Quality of education as it relates to climate change adaptation illustrates qualitative change along the following two dimensions: (i) qualitative change in the classroom/learning environment and (ii) qualitative change in the flow of students/learners. The realisation and effectiveness of these dimensions rest on the dynamic nature of the teaching learning environments.

Quality education should have at its focus the realisation of quality teaching/instruction. Quality teaching and quality learning outcomes are indications of quality of the functioning of the various related units of the Ministry or Department of Education. Therefore, quality teaching and learning then provide the groundwork for quality management and administration.

From a broader conceptual sense, one of the purposes of education is the development of socialisation processes, i.e. new generations take ownership of the contents of culture and acquire the competencies necessary to actively participate in society. Culture is the experience accumulated by different social groups that have generated collective solutions for facing their difficulties and that encompass aspects such as conceptual and interpretive frameworks of reality, beliefs, ways of life, languages and others (Sasaki and Biro 2017). Therefore, it is necessary for further research exploring pathways on the utilisation of education for the development of a sustainable climate change adaptation culture in West Africa in the future.

The curriculum is an essential element in the development of vital climate change adaptation education pathways. The school curriculum represents a collection of those contents of culture that, at a given time, are considered important for fostering the development of persons and for achieving the purposes sought by education. According to William (2013) the word curriculum has no generally agreed meaning other than it describes the educational experiences planned for learners in educational institutions. Ralph Tyler (1949) proposed that the curriculum should be viewed to an end, rather than an end in itself. He proposed four questions that must be answered when developing any curriculum:

1. What educational purposes should the school/institution seek to attain?
2. What educational experiences can be provided that are likely to attain these purposes?
3. How can these educational experiences be effectively organised?
4. How can we determine whether these purposes are being attained (Tyler 1949 p. 1)?

The curriculum should be viewed as a creative process based on four key elements including objectives, evaluation, knowledge and school learning experience with the belief that the elements relate with each other, so that a change in one element will lead to a change in the others (Kerr, 1968, Williams, 2013). The world is climate changing rapidly, and although we know that the changes will continue, it is extremely difficult to predict the precise impact such changes will have on humanity (Hamel, 2001). Therefore, the education system must prepare learners to respond adequately to these changes (Williams, 2013). Curricular frameworks and programmes of study structure the acquisition of knowledge, skills and attitudes the society wish to through different areas of learning (UNESCO, 2007). Therefore, it is important to mainstream climate change adaption education for sustainable development throughout the education system.

In summary, education is a viable vehicle to contribute to CCA and sustainable development. Therefore, protecting and utilising education infrastructure, specifically the social and material resources on which education depends, can reduce vulnerability and build resilience among communities. Efforts to improving education standards, including measures to enhance literacy, school and academic attainment, can enhance the adaptive capacity of the society, and research-based climate change adaptation learning support can hasten social and policy change by amplifying learning before and during adaptive decision-making.

Youth, TVET and Adult Education in Africa

This section of the book presents a perspective on the situation of TVET in Africa, the challenges faced by the sector and strategies for addressing the emerging challenges.

Some countries in Africa are exploring ways to implement strategies that enhance both the employment prospects of young women and men and the productivity of enterprises (Brewer & Paul, 2015). Quality demand-driven TVET and skills for employment development are essential tools for equipping young people with the necessary skills they will need to create sustainable livelihoods and participate actively in the development of society. Core skills are gaining increased importance and significance in the TVET and across the skills sectors.

The prospects of young women and men and the productivity of enterprises are gaining new currency in education and sustainable employment in Africa. According to Brewer and Paul (2015), new emerging concepts in public management reforms that included new governance structures are giving institutions more autonomy and businesses more say at local and national levels; sectoral bodies; competency-based curricula; national qualifications frameworks; and outcomes-based funding borrowed from the old Commonwealth toolkit. This new perception of TVET is based on the human capital theory, which argues that it is what is learnt that is the mechanism for enhanced efficiency and production, wages and economic development (McGrath et al., 2019).

In a review of evidence on labour market and entrepreneurship programmes (McGrath et al., 2019) argued that Africa's employment programmes show minimal positive effects. The reviewers assumed that TVET interventions were poor value for money. Similar evaluations of active labour market policies also conclude that vocational programmes were largely ineffective. These reviews were faulty because (a) the focus was on a small sample of a specific type of intervention rather than a more comprehensive sample of TVET initiative across Africa, (b) externally funded and (c) short-duration vocational programmes in settings where there is an unemployment crisis and not on mainstream TVET programmes (McGrath et al., 2019). Another criticism of TVET offering in Africa is that they were deemed unproductive because they were locked into a narrow human capital understanding that is too dependent on notions of individuals as acting out of narrow economic rationality and downplays the broader structural issues impacting the overall sector. Few countries in Africa have initiatives to develop the technical and vocational skills required for the climate change transition in their national adaptation plans. A further fewer countries have laid out strategies specific to skills development, technological advancements and creating a vocationally focused workforce, recognising the emerging demand for the skills required by a green economy. Therefore, a conceptual analysis of the existing policies is needed, and a shift in actions and strategies for climate action required to deal with the gap in policy and processes (UNESCO, 2021).

Therefore, to enhance the perspective and relevance of the TVET sector to the achievement of the Sustainable Development Goals in Africa, consideration must be focused on more expansive areas such as teaching and learning, curriculum development, pedagogy and student support (Needham, 2018; Papier & McBride 2018). It is also essential that consideration is focused on using educational technologies within TVET settings to enhance overall delivery and access of training (Rotich et al., 2016). McGrath and colleagues (2019) highlight the issues regarding languages of instruction within the setting. They maintain that due to the low levels of student competence in these languages, they cannot perform efficiently in these TVET programmes. The issue of instructional language was addressed through the training of instruction sessions.

An essential outcome of the YEEP was to address the archaic management systems in practice in TVET programmes to more modern participatory management practices, thus utilising local languages for instruction. The achievement of this outcome led to the improvement in management systems (McGrath et al., 2019). During the project, efforts were made to enhance the partnership of the colleges and TVET programmes with other institutions involved in youth development and TVET programming (Ayentimi et al., 2018), including interventions to improve college-to-work transitions through internship and apprenticeship initiatives with private and public sectors (McGrath et al., 2019). In summary, the following are some critical issues to be considered when implementing TVET in Africa.

Vocational Knowledge

The analysis of TVET in Africa brings to the forefront some critical issues regarding the relevance of the knowledge and content that students should deliver in TVET settings. The problems of what kinds of subject knowledge and practical skills should be taught to learners at which levels are critical. There must be a balance between literature to be taught and the relationship between theory and practice. According to McGrath et al. (2019), this highlighted the importance of discipline-based curricula as the basis for meaningful work and the more comprehensive broader framework within which practical skills and knowledge are located.

There is a mismatch between the knowledge and skill required by the market and those taught by education and training institutions (Carter, 2008; UNEP, 2013). The uncertainties relating to the impact of climate change (UNDP, 2021) provide a significant challenge for policymakers and education officials in imagining new vocational provisions for the future. The literature points to the need to get beyond crude technical approaches to what skills appear to be needed at the surface level and consider what knowledge and skills are required for transformative TVET (McGrath et al., 2019). This involved understanding the knowledge basis of new programmes, qualifications and occupations necessary for the sector and its link with the achievement of the broader sustainable development goals (Simmons, 2021).

Critical Capabilities Approach

Throughout the implementation period of YEEP, new theoretical approaches to TVET and sustainable development were emerging, utilising the human development and capabilities approach (Sen, 1999). The critical capabilities approach (CCA) addresses equally the emerging mismatch and inequality in skills for employment and development and how to shift from a narrow focus of addressing current employability and production to addressing future employment needs (McGrath et al., 2019).

The critical capabilities approach (CCA) is forward-looking and more sustainable in its construct. It moves further than the human capital approach by developing a much robust account of the organisation. It has also extended past early work on capabilities and education for sustainability by insisting on the importance of structure and power. The model is guided by the political economy of skills tradition, by feminist theory. The methodology emphasises the need to consider young people's voices in articulating their perspectives and aspirations for decent work and wellbeing (McGrath et al., 2019).

TVET for Community Development

This notion of TVET evolved from adult and community education ingrained in the early years and grounded in participatory action research (Tukundane et al., 2015). It is also rooted in valuing and acting upon local knowledge, including those of rural people and the youth (Kanyandago, 2010), in ways that are paralleled to the notion of achieving sustainable development. With CCA, this shift of focus to local socioeconomic contexts, attitudes and aspirations appears crucial for a transformative approach to TVET in Africa.

Skills for Sustainable Development

Due to the severity of climate change and other development calamities, there is a need to tackle environmental concerns while also addressing societal and economic considerations. Therefore, training youth in natural resource management-orientated TVET can be viewed as early preparation of youth in 'green' skills. Thus, the need for integrating green skills into more comprehensive national skills systems is more pressing in the face of severe environmental challenges. The focus in TVET in the future should be on 'greening the economy; sustainable energy development; sustainable consumption and production; integrated waste management; and sustainable land management' (McGrath et al., 2019; Poschen, 2015; ILO, 2010; Simmons, 2021).

Innovations to Assess and Enhance the Institutional Capacity of Youth Agencies

Building on the work in the book entitled *Impacts of Climate Change on Young People in Small Island Communities* on 'top-down' and 'bottom-up' approaches and human capital framework, the author explored the effectiveness of other innovations, including (a) cultural web, (b) information technologies and information system (IT/IS) and (c) business process reengineering as strategic tools to assess the culture and management of development institutions and enhancing the capacity and efficiency of these institutions.

(a) Cultural Web

An organisation's culture is locked into the beliefs and norms of every individual within that organisation; therefore, if left on its own, it would be exceedingly difficult to change. Armstrong (1999) supports this view. He argues that if there was appropriate and effective culture within an organisation, then it would be desirable to take steps to support it, whereas, if the culture is inappropriate, all attempts should be made to determine what needs to be changed and to develop and implement a plan for that change.

Management needs to have a clear understanding of the culture of that organisation to implement successful changes within an organisation. Therefore, an organisation can be captured by its culture, which is the product of history and the institutional capacity to implement successful changes within and across organisations (Johnson et al., 2017; Johnson & Scholes, 1988; 2002).

The cultural web is defined as representing the *taken for granted* assumptions or paradigm of an organisation and the physical manifestation of organisational culture. The concept of the cultural web is useful in that it can be used to analyse the way in which an organisation behaves. This is a particularly useful tool for analysing the culture within the National Youth Commission of Sierra Leone and will provide a clearer understanding of the *taken for granted* assumptions in the organisation.

In the cultural web model, six cultural components, including power structure, organisational culture, control systems, stories, control systems, rituals and routines and symbols, are utilised to assess the culture of the institutions. The cultural web is defined as the demonstration of the organisational culture and employs behaviours, reflecting the collective experience of the organisation (Cooper et al., 2018). The cultural web framework has its critics. Researchers believe that the cultural web models and the identification of the six elements may not be enough to understand and measure the organisational culture (Cooper et al., 2018) due to the complexities associated with modern organisations. Other researchers argue that the culture of an organisation is tangible, therefore making it impossible to measure its everyday operations and the way some organisations are managing certain departments (Cooper et al. 2018, Sun, 2008).

(b) Information System and Information Technologies (IS/IT)

The capability of a society is highly affected by its accumulation of knowledge and its effectiveness in using this knowledge in sustainable ways to create new knowledge and skills. As technology evolves and expands, so too is the emphasis on its use in organisational changes. Organisations are acquiring the most advanced technologies to give them the competitive advantage. The leading-edge focus in institutions today for technology application is now around decision support and competitive advantage. Leading organisations are using information technology not just to support their work at the operational level but more importantly to incorporate IT at the strategic level. Strategic information system (SIS) is defined as systems that are designed to bring competitive advantage or have resulted in competitive edge to the organisation (Hickman, 2002). SIS is used for various purposes including (a) support strategic decision-making, (b) shaping and/or defining the organisation's competitive strategy and (c) for a combination of the above (Cavaye & Cragg, 1993).

Research shows that investment in information technology has caused varying levels of changes in institutions (Ward & Elvin, 1999). However, institutions are not equipped with the knowledge and skills of how to use ICT for strategic decision-making (Hickman, 2002). The author uses IT/IS to assess how IT is used for making strategic decisions to facilitate organisational change.

> Keeping up with new information and knowing how to use it are 'mission critical' activities to businesses and individuals alike in a market where competition is no longer characterised by the big beating up the small, but rather by the fast running past the slow. (Ruttenbur et al., 2000, p15).

The main purpose for implementing IS/IT during the YEEP was to change the business and the organisation in some significant ways. It is extremely difficult to effectively evaluate the advantages of such application to the organisation because only 25% of organisations regularly review the benefits delivered at the end of a project. Therefore, it is important for organisation to put the necessary mechanisms in place to ensure that benefits are analysed and assessed because such assessment is utilised to determine whether the IS/IT investment/application meets the need of the organisation (Cavaye & Cragg, 1993, O'Brien, 2000).

IS/IT and E-Learning

IS/IT utilised e-learning mechanism to give tertiary and TVET institutions its competitive edge during the implementation of the YEEP. The project delivered skills training for approximately three times the number of youths that were proposed by the government in the YEEP. The success of the YEEP supported the view that traditional learning style was delaying the presentation of knowledge in the modern era (Bhalalusesa et al., 2013); therefore, in essence, the traditional learning methods are not suitable for today's fast-growing knowledge-driven economies (Ruttenbur et al., 2000; Rosenberg, 2000). The strategic use of IT/IS and the Internet applications can be included as a successful way to blended learning, e-learning, and m-learning approaches which are part and parcel of the digital education age (Elzawi et at., 2013).

Training and education are defined as 'giving people the information and skills they need to compete effectively in the marketplace' (Ruttenbur et al., 2000, p15). Based on this definition of education, education and training institutions need to understand this definition to enable them to provide ways of certifying and equipping individuals with enough skills to cope with the demands of modern society (Thuraya, 2015).

E-learning is one of the educational vehicles that has emerged from the growth and promotion of information and telecommunication technology (Thuraya, 2015). The principal outcome of e-learning is basically learning which involves the usage of any electronic device, from computers to mobile phones, and which might, or might not, involve the usage of the Internet (web sites+ other applications) or an intranet (local network system). E-learning could be presented through several resources, for example, computer software and Internet websites. Moreover, other applications have been developed specifically for e-learning, such as virtual learning environments (VLEs), which provide the user or the learner with numerous facilities to access the learning materials and communicate with lecturers and or trainers and their peers. The advantage of the VLEs is that they provide flexible access to learning, as the learning can be accessed from anywhere and at any time (Adam & Healy, 2000; Rosenberg, 2000).

Business Process Reengineering

Business process reengineering (BPR) is defined as a radical reconsideration of a business process to achieve dramatic improvement in cost, quality, service and speed performance (Heflo.com/blog/bpm/business-process-reengineering-examples). BPR was used by the Project Implementation team to enhance the overall operational processes of the NAYOM to implement the YEEP in Sierra Leone. The business process reengineering (BPR) is viewed as a methodology for seeking major break-through in the broad project management processes. It is a collection of one or more types of work activities that adds value to the input for customers and is concerned with studying the work of a company/organisation from a process standpoint to improve its work method (Dilworth, 1996).

The following consideration should be given when implementing BPR to manage organisational change: (a) change focus of putting emphasis on the management of the institutions to that of the customers, (b) it is pertinent that the management of the implementing organisation gives power to the team that is responsible for the implementation of the project, (c) imperative for the institution to focus on the result anticipated by the change management process, (d) vital to provide leadership for the implementation of the change process by providing opportunities for teaching and learning to take place within the institutions, and (e) ensure that the operational processes are simple rather than being complicated and intricated (Heflo.com/blog/bpm/business-process-reengineering-examples).

BPR is a combination and blends of operation management tools such as *just-in-time concept, process flow-charting, critical examinations in method study, operation new-work management and customer focus operation* (https://www.Heflo.com/blog/bpm/business-process-reengineering-examples). The considerations expressed above were used by the implementing team to determine the appropriate BPR tools for implementing the YEEP. The implementation team also (a) looked at the future and (b) identified the goals and purpose of using the BPR tools and (c) the mission of the youth institutions when implementing the BPR tools. The YEEP team did not waste any time in discarding tools, methodologies and/or processes that were not working and coming with new and effective workable tools/models.

The role of the customer is central when implementing BPR. Therefore, it is essential to reiterate that when implementing BPR, employees need to realise that they are not working for their bosses on keeping pace with targets but simply for the customers (Hammer & Champy, 1993). They expanded on the above view by adding that power needs to be given to the small teams who have the capabilities, competence and skill to respond quickly to issues from a customer focus perspective.

During the YEEP implementation process, the project team further delved into the use of BRR and critical path to enhance efficiency in the overall management and operation of the project process. This was based on the belief that the operations of the project should be organised around the total process of a business that adds value to its customers, rather than the activities which perform the stages of the value-added activity (Slack et al., 1998). Beardwell and Holden (1997) contend that the principle may not seem particularly revolutionary, but the uniqueness lies in how the key principles are instituted, the holistic nature of the package and the benefits that the key principle is said to deliver.

The benefits of BPR are that it provides the opportunity to integrate a whole set of practices from making organisation's flatter, shifting from training to education to learning, incorporating pay and IT systems, adds values and creating an environment where managers coach rather than supervise (Hammer & Champy, 1993). It allows managers to see the world as dangerous and unstable, and it is economy as vulnerable and subjected to rapid shifts and tensions. Although there are numerous benefits to be realised by the organisation, it is believed that 80% are still not implementing BPR (Watson, 1994 in Beardwell and Holden (ibid).

Summary of Chapter

The purpose of the Review of Literature in this chapter is to build on the literature surrounding the earlier book on the *Impacts of Climate Change on Youth in Small Island Communities* by exploring existing and other new bodies of literature required to support the implementation of the Youth Climate Change Adaptation Framework to support the implementation of the Youth Employment and Entrepreneurship (YEEP) in Sierra Leone, West Africa.

The review was conducted to further explore the applicability and relevance of the top-down, bottom-up, and other critical approaches to the successful implementation of the Youth Climate Change Adaptation Framework in West Africa. The findings from the review of the literature show that there is an acceptable body of knowledge on vulnerability and adaptation as they relate to skills for employment, curriculum development and training in relationship to building the capacity of youth to adapt to the impact of climate change and other calamities. The chapter explores a range of methodologies relevant for training instructors/lecturers and youth at tertiary and TVET institutions. These methodologies were applicable in training young people in climate change adaptation and skills development for employment skills via entrepreneurship training in West Africa.

The chapter also focuses on strategies and processes for developing climate change adaptation education for sustainable development curricula in skills for employment through entrepreneurship. The chapter explores the relevance of using strategic management tools such as (a) business process reengineering, (b) information technologies systems and (c) cultural web to (i) assess the gaps in the capacity of youth organisations and (ii) devise strategies to enhance the capacity of these institutions to deliver quality education and sustainable livelihood initiatives.

Due to the peculiarities and complexities of utilising one approach over the other, the author was guided by the body of literature towards utilising a combination of approaches to enhance the curricula, training modules design and implementation and capacity development of youth development institutions. Academic rigour was applied by exploring appropriate human capital strategies to enhance the capacity and resilience of young people to adapt to the impact of climate change, building sustainable livelihoods via entrepreneurship now and in the future, and contribute to the development of national economic planning processes.

References

Adam, F., & Healy, M. (2000). *A practical guide to postgraduate research in the business area*. UK.

Adger, N. W., Arnell, N. W., & Tompkins, E. L. (2005). Successful adaptation to climate change across scales. *Global Environmental Change, 15*(2), 77–86. Available at: http://www.sciencedirect.com/science/article/pii/S0959378004000901 Accessed July 11, 2014

Armstrong, M. (1999). *How to be an even better manager* (5th ed.). Kogan Page Limited, Pentonville Road.

Ayentimi, D., Burgess, J., & Dayaram, K. (2018). Skilled labour shortage. *Human Resource Development International, 21*(5), 406–424. https://doi.org/10.1080/13678868.2018.1447881

Arnell, N. W., Lowe, J. A., Brown, S., Lincke, D., & Price, J. T. (2016). *The global impacts of climate change under a 1.5oC pathway: Supplement to assessment of impacts under 2, 3 and 4oC pathways*. Report from AVOID2 project to the Committee on Climate Change.

Beardwell, I., & Holden, L. (1997). *Human resource management: A contemporary perspective* (2nd ed.). Pearson Education Limited, Prentice Hall Publishing.

Benjamin, L. (2009). Climate change and the Caribbean Small Island states: The state of play. *International Journal of the Bahamian Studies, 16*, 78–91. Available at: http://researchjournal.cob.edu.bs

Bhalalusesa, R., Lukwaro, E. E., & Clemence, M. (2013). Challenges of using e-learning management systems faced by the academic staff in distance-based institutions from developing countries: A case study of the Open University of Tanzania. *Huria Journal of OUT, 14*, 89–110.

Bradshaw, Corey, J. A., Ehrlich, P. R., Beattie, A., Ceballos, G., Crist, E., Diamond, J., Dirzo, R., Ehlich, A. H., Rodolfo, M. E., Pyke, G., Raven, P. H., Ripple, W. J., Saltre, F., Turnbull, C., Wackernagel, M., & Blumstein, D. T. (2021, January 31). Underestimating the challenges of avoiding a ghastly future. *Frontiers in Conservation Science, 1*, 1–10. https://doi.org/10.3389/fcosc.2020.615419

Brewer, L., & Paul, C. (2015). *Integrating core work skills into TVET systems: six country case studies/Laura Brewer, Paul Comyn; International Labour Office, Skills and Employability Branch, Employment Policy Department* (p. 2015). ILO.

Bueno, R., Herzfeld, C., Stanton, E. A., & Ackerman, F. (2008). *The Caribbean and climate change: The coast of inaction*. Tufts University.

Burton, I., Huq, S., Lim, B., Pilifosova, O., & Schipper, E. L. (2002). From impact assessment to adaptation priorities: The shaping of adaptation policy. *Climate Policy, 2*(23), 145–159.

Carter, R. (2008). *Caribbean youth: An integrated literature review*. Bridgetown.

Cavaye, A. L. M., & Cragg, P. M. (1993). Strategic information system research: A review and research framework. *Journal of Strategic Information System, 2*(2), 125–137.

Commonwealth Secretariat. (2007). *The commonwealth plan of action on youth empowerment 20072015*. Commonwealth Youth Programme, Commonwealth Secretariat..

Cordero, E. C., Centeno, D., & Todd, A. M. (2020). The role of climate change education on individual lifetime carbon emissions. *PLoS One, 15*(2), e0206266. https://doi.org/10.1371/journal.pone.0206266

Cooper, M., Collins, M., Bernard, R., Schwann, S., & Knox, R. (2018). Criterion-related validity of the cultural web when assessing safety culture. *Safety Science, 111*(1), 49–66.

Dilworth, J. B. (1996). *Operations management* (2nd ed.). McGraw-Hill Publishing.

Fieser, E. (2017). *Irma is the Caribbean's Most expensive storm at $10 billion and counting*. Bloomberg Business. Available at: https://www.bloomberg.com/news/articles/2017-0908/most-expensivestorm-ever-in-caribbean-did-10-billion-in-damage

Gomez, O. (2013). *Climate change and migration: A review of the literature*. International Institute of the Social Studies, The Hague (Erasmus University Rotterdam), Rotterdam.

Government of Sierra Leone. (2014). *A Blue Print for Youth Development: Sierra Leone's National Youth Programme 2014–2018, Ministry of Youth Affairs*. Freetown, Sierra Leone.

Government of Sierra Leone. (2015). *Population and housing census*. Government of Sierra Leone.

Government of Sierra Leone. (2019). *Youth skills development for entrepreneurship and employment, 2019*. National Youth Commission (NAYCOM).

Hamel, G. (2001). *Leading the revolution*. Harvard Business School Press.

Hammer, M., & Champy, J. (1993). *Reengineering the corporation: A manifesto for business revolution*. Nicholas Brealey Publishing.

Hernandez-Delgado, E.A. (2015). *The emerging threats of climate change on tropical coastal ecosystem service, public health, local economies and livelihoods sustainability of small islands: Cumulative Impacts and Synergies*. Marine Pollution Bulletin, University of Puerto Rico, Centre for Applies Tropical Ecology and Conservation, Coral Reef Research Group Puerto Rico, USA.

Hjem, H., & Hull, C. (1982). Implementation research as empirical constitutionalism. *European Journal of Political Research, 10*, 105–116.

Hickman, S. (2002). *Handout on strategic information system planning (SISP) – Information technology and business, masters in business administration*. University of Greenwich.

ILO. (2010, November). *A skilled workforce for sustainable growth: A G20 training strategy*. International Labour Office.

ILO. (2013). *Sustainable development, decent work and green jobs*. International Labour Conference, 102nd Session, International Labour Office, Geneva.

IPCC. (2001). Climate change 2001: Impacts, adaptation, and vulnerability. In *Contribution of Working Group II to the third assessment report of the Intergovernmental Panel on Climate Change*. Cambridge University Press.

IPCC. (2014a). Climate change (2014): Synthesis report. In *Contribution of working groups I, II and III to the fifth assessment report of the intergovernmental panel on climate change* (Vol. 151). IPCC.

IPCC. (2014b). Summary for policymakers. In: Climate change impacts, adaptation and vulnerability. In C. B. Barros, V. R. Dokken, D. Mach, K. J. Mastrandrea, M. D. Bilir, T. E. Chatterjee, M. Ebi, & K. L. Estrada (Eds.), *Contributions of the working group II to the fifth assessment report*. Field.

IPCC. (2014c). Climate change: Synthesis report. In *Contribution of working group 1, 11, 111 to the fifth assessment report of the intergovernmental panel on climate change* (p. 151). IPCC.

IPCC. (2015). *Report of the Conference of the Parties on its twenty-first session, held in Paris from 30 November to 13 December 2015*. Addendum. Part one & two: Action taken by the Conference of the Parties at its twenty-first session. https://unfccc.int/documents/9097#beg

Jaja, J., Dawson, J., & Gaudet, J. (2016). *Using social network analysis to examine the role that institutional integration plays in community based adaptive capacity to climate change in Caribbean small island communities*. Department of Geography, Environment and Geomatic, University of Ottawa, Canada: Routledge, Local Environment.

Johnson, G., & Scholes, K. (1988). *Exploring corporate strategy*. Prentice Hall.

Johnson, G., & Scholes, K. (2002). *Exploring corporate strategy* (6th ed.). Prentice Hall.

Johnson, G., Scholes, K., & Whittington, R. (2017). *Exploring corporate strategy: Text & cases* (11th ed.). Pearson Education Limited.

Kanyandago, P. (2010). Revaluing the African endogenous education system for community-based learning. In J. Zeelen, J. van der Linden, D. Nampota, & M. Ngabirano (Eds.), *The burden of educational exclusion* (pp. 99–114). Sense.

Matland, R. (1995). Synthesising the implementation literature: The ambiguity-conflict model of policy

implementation. *Journal of Public Administration Research and Theory, 5*(2), 145–174.
Marshall, P. A., Johnson, J. E. (2007). *The great barrier reef and climate change: Vulnerability and management implications*. Climate change and the Great barrier reef Marine Park Authority and the Australian Greenhouse Office, Australia, pp. 774–801.
McGrath, S., Ramsarup, P., Zeelen, J., Wedekind, V., Allais, S., Lotz-Sisitka, H., Monk, D., Openjuru, G., & Russon, J.-A. (2019). Vocational education and training for African development: A literature review. *Journal of Vocational Education and Training, 72*(4), 465–487.
McNamara, K. E., & Buggy, L. (2016). Community based climate change adaptation: A review of academic literature. *Local Environment*. https://doi.org/10.1080/13549839.2016.1216954
Mills, E. (2005). Insurance in climate of change. *Science, 309*, 1040–1044.
Nathaniel, O., & Tetteh, A. J. (2017, December). The impact of climate change in Africa: A review of literature. *Journal of International Academic Research for Multidisciplinary Impact Factors, 5*(4483) ISSN: 2320-5083.
Needham, S. (2018). Student support structures for transitioning from vocational to university education. In S. McGrath, M. Mulder, J. Papier, & R. Suart (Eds.), *Handbook of vocational education and training* (pp. 1583–1594). Springer.
Nurse, L. A., McLean, R. F., Agard, J., Briguglio, L. P., Duvat-Magnan, V., Pelesikoti, N., Tompkins, E., & Webb, A. (2014). Small islands. In V. R. Barros, C. B. Field, D. J. Dokken, M. D. Mastrandrea, K. J. Mach, T. E. Bilir, M. Chatterjee, K. L. Ebi, Y. O. Estrada, R. C. Genova, B. Girma, E. S. Kissel, A. N. Levy, S. MacCracken, P. R. Mastrandrea, & L. L. White (Eds.), *Climate change 2014: Impacts, adaptation, and vulnerability. Part B: Regional Aspects*. Contribution of Working Group II to the Fifth Assessment Report of the Intergovernmental Panel on Climate Change Cambridge University Press, Cambridge, United Kingdom and New York, NY, USA, pp. 1613–54.
O'Brien, J. A. (2000). *Management information systems; managing information technology in the E-business Enterprise* (5th ed.). Mc Graw-Hill Publishing.
OECD. (2013). *The nature of policy change and implementation: A review of different theoretical approaches*. Organisation for Economic Cooperation and Development.
Olmos, S. (2001, July). *Vulnerability and adaptation to climate change: Concepts, issues, assessment and methods*. Foundation paper, Climate Change Knowledge Network.
Papier, J., & McBride, T. (2018). Systematizing student support services in TVET colleges. In S. McGrath, M. Mulder, J. Papier, & R. Suart (Eds.), *Handbook of vocational education and training* (pp. 1595–1610). Springer.
Pereznieto, P., Gbedemah, C., Rosenem, G., Haper, C., & Jones, N. (2011). *Youth vulnerability and adaptation: Exploring the impact of macro-level shocks on youth 3F crisis and climate changes*. Overseas Development Institution.
Poschen, P. (2015). *Decent work, green jobs and the sustainable economy: Solution for climate change development*. International Labour Organisation.
Paytan, A., Catherine, H., Emily, W., Pedemonte, S., & Esra, M. (2017). Education or climate change adaptation and resilience: ASLO global outreach initiative. *Bulletin of Limnology and Oceanography, 26*(3). https://doi.org/10.1002/lob.10185
Qaisrani, A. (2015). *Connecting the dots: linking climate change resilience to human capital*. SDPI & PRISE.
Quisumbing, L. R. (2001). *The importance of values education for TVET and its economic and human resource development*. UNESCO Asia Pacific Conference, Adelaide. yooyahcloud.com.
Ripple, W. J., et al. (2019). World scientist warning of a climate emergency. *Bio Science, 70*(1), 100.
Rosenberg, M. (2000). *E-learning: Strategies for delivering knowledge in the digital age*. McGrawHill.
Ruttenbur, B., Spickler, G., & Luri, S. (2000). *E-learning the engine of the knowledge age*. Morgan Keegan & Company. Retrieved November 27, 2013. From: http://intemettime.com/itimegroup/morgankeegan.pdf.
Rotich, A., Kosgey, J., & Kimutai, K. (2016). IT best practices for implementing E-learning: The case of Rift Valley Technical Training Institute. *Africa Journal of Technical and Vocational Education and Training, 1*, 93–100. [Google Scholar].
Sasaki, T., & Biro, D. (2017). Cumulative culture can emerge from collective intelligence in animal groups. *Nature Communications, 8*, 15049.
Schleussner, C. F., Lissen, T. K., Fischer, E. M., Wohland, J., Perrette, M., Golly, A., Rogelj, J., Childers, K., Schewe, J., Frieler, K., Mengel, M., Hare, W., & Schaeffer, M. (2016). Different climate impact for policy-relevant limits to global warming: The case of 1.5 °C and 2 °C. *Earth System Dynamics, 7*, 327–351.
Sen, A. (1999). *Development as freedom*. Oxford University Press.
Simpson, M. C., Scott, D., Harrison, M., Simm, R., Silver, N., O'Keeffe, E., Harrison, S., Taylor, M., Lizcano, G., Rutty, M., Stager, H., Oldham, J., Wilson, M., New, M., Clarke, J., Day, O. J., Fields, N., Georges, J., Waithe, R., & McSharry, P. (2010). *Quantification and magnitude of losses and damages resulting from the impact of climate change: Modelling the transformational impacts and costs of sea-level rise in the Caribbean*. United national Development Programme (UNDP).
Simmons, A. (2006). Youth development index; commonwealth. *Journal on Youth and Development, UNISA, South Africa, 4*(2), 100–113.
Simmons, A. (2021). *Impact of climate change on youth in small island communities* (Sustainable Development Goals Series, Climate Change) (Vol. 1). Springer Nature.
Slack, N., Chambers, S., Harland, C., Harrison, A., & Johnston, R. (1998). *Operations Management* (2nd ed.). Pearson Education Limited.
Stochowiak, S., Robles, L., Habtermanion, E., & Maltry, M. (2016). *Beyond win: Pathway for policy implementation*. ORS Impact/The Atlas Learning Project.

Suggett, D. (2011). *The implementation challenge: Strategy is only as good as execution*. State Services Authority Occupational paper No. 15.

Sun, S. (2008). Organizational culture and its themes. *International Journal of Business and Management, 3*(12), 137–141.

Sustainable Development Solutions Network – Youth. (2020). *Youth solution report - 4th edition, sustainable development solutions network – Youth*. UNDP. https://www.undp.org/content/undp/en/home/librarypage/democratic-governance/youth-solutions-report-2020.html

Takewira, I., & Simmons, A. (2004). *Youth development at the centre of the Commonwealth agenda. Commonwealth Youth Programme (CYP)*. Forum 21, European Journal on Youth Policy, No 2.12/2004 pp. 82–91.

The Economist. (2017). *Paradise lost How Hurricane Irma will change the Caribbean: The region must adapt to climate change, not simply rebuild*. Print edition | The Americas, Sep 14th 2017.

Thuraya, K. (2015). *Improving the effectiveness of E-learning improvement in the school of engineering at Tripoli University*. Doctoral thesis, University of Huddersfield.

Tukundane, C., Minnaert, A., Zeelen, J., & Kanyandago, P. (2015). Building vocational skills for marginalised youth in Uganda. *International Journal of Educational Development, 40*, 134–144.

Tyler, R. W. (1949). *Basic principles of curriculum and instruction*. University of Chicago Press.

Kerr, J. F. (1968). The problem of curriculum reform. In J. F. Kerr (Ed.), *Changing the curriculum* (pp. 33–38). University of London Press.

United Nation. (1992). United Nations Conference on Environment & Development, Sustainable Development, Rio de Janerio, Brazil, 3 to 14 June 1992 AGENDA 21.

United Nations. (2014). *Emerging issues for small island developing states. Result of UNEP Foresight process*. UNEP.

United Nations. (1992). *NGO Final Report on the proceedings of the Conference on the Environment and Development* (Earther Summit Bulletin, Vol 2, No 12). Island Press and International Institute on Sustainable Development. Tuesday 16th June 1992.

UNDP. (2021). *World's largest survey of public opinion on climate change: A majority of people call for wide-ranging action*. United Nations Development Programmes.

UNDP. (2015). *Human development report*. Work for Human Development, United National Development Programme (UNDP).

United Nations Environment Programmes. (2013). *Identifying emerging issues from the perspective of Small Island Developing States (SIDS)*. Expert Group Meeting Co-hosted by the United Nations Environment Program (UNEP) and the United Nations Department of Economic and Social Affairs (DESA), 14–16 May 2013, Cambridge, United Kingdom.

UNEP. (2021). *GEO-6 for Youth*. UN Environment Programme. Kenya, Nairobi.

UNESCO. (2007). *Quality Education for All: A Human Rights Issue*. Intergovernmental Meeting of the Regional Education Project for Latin America and the Caribbean, 2 nd Buenos Aires 2007 (11). ISBN 978–956-8302-70-2.

UNESCO. (2014). *Shaping the future we want: UN decade of education for sustainable development (2004–2014)*. DESD Monitoring and Evaluation.

UNESCO. (2015). *Not just hot air: putting Climate Change education into practice*. Education sector, Paris, France.

UNESCO. (2021). *Skills development and climate change action plans Enhancing TVET's contribution*: UNESCO & UNESCO-UNEVOC International Centre for Technical and Vocational Education and Training UN Campus Platz der Vereinten Nationen 1 53113 Bonn Germany: ISBN: 978-92-3-100438-4. https://unevoc.unesco.org/pub/skills_development_and_climate_change_action_plans.pdf

UNFCCC LEG. (2017). Draft recommendations developed by the Adaptation Committee and the Least Developed Countries Expert Group to address decision 1/CP.21, paragraph 41 and, in collaboration with the Standing Committee on Finance, to address decision 1/CP.21, paragraph 45, Adaptation Committee Least Developed Countries Expert Group (LEG), UNFCCC-LEG. https://unfccc.int/topics/adaptation-and-resilience/groups-committees/adaptation-committee/joint-ac-and-leg-mandates-in-support-of-the-paris-agreement

UNTAD. (2018). *Economic development in Africa report 2018: Migration for structural transformation*. United Nations publication issued by the United Nations Conference on Trade and Development.

Ward, J., & Elvin, R. (1999). *A New framework for managing IT- enabled business change* (pp. 197–221). School of Management.

Watson, T. (1994). *In search of management: Culture, chaos and control in management work*. Routledge.

Williams, D. (2013). *Redesigning Schooling – 3: Principled curriculum design: Redesigning School, the Campaign schoolish vision for education*. SSAT (The School Network) LTD.

World Bank. (2007). *World development reports 2007: Development and the next generation*. World Bank.

World Bank. (2013, June). *Quality education count for skills and growth*. Caribbean Knowledge Forum Series: Caribbean Growth Forum, World Bank LAC.

3 Application of Climate Change Adaptation Framework in Sierra Leone and Its Relevance to Enhancing Public-Private Partnership Policies and Practice in TVET Development

Abstract

The chapter focuses on the implementation of the CCA Framework for youth development. The chapter is divided into three sections. Section 1 discusses the methodological processes involved in implementing the four stages of the framework, including the challenges and successes faced by the consortium in Sierra Leone, West Africa. The second section of the chapter is divided into two subsections: (a) the Public-Private Partnership (PPP) framework and (b) the YEEP Pilot Model Framework. The PPP framework captures a range of systems to share skills development responsibility among TVET stakeholders and employers. The YEEP pilot model framework is a conceptual document highlighting the processes, structures and mechanisms required by the stakeholders to implement youth development programmes in developing countries. Together, the frameworks link TVET training and employers, and engage the private sector in skills development.

Keywords

Climate change · Sustainable development · Youth development · Entrepreneurship · Employment · Climate change adaptation · Africa · Biodiversity · Technical vocational education training · Conceptual framework · Public-private partnership · Memorandum of understanding

Introduction

Like other developing countries in Africa, Sierra Leone is constantly faced with shortages in financial resources to support its education sectors. The Public-Private Partnership (PPP) framework developed by the consortium encourages private sector agencies to partner with the government by providing funds to implement the educational programmes within the TVET sector and bolster participants' job readiness (Xiong et al., 2018). The PPP framework is designed to capture a range of public policies, funding systems and curriculum frameworks with shared participation in skills development among TVET stakeholders and employers. The PPP policy paper is supported by the YEEP Pilot Model Framework, which provides an overview of the stages, procedures and processes for consideration and actions when setting up YEEP and/or similar youth sustainable development livelihood project.

The chapter is divided into three sections and presents the nuts and bolts for developing the key policy documents and frameworks that evolved out of the implementation of YEEP. Section 1 discusses the methodological processes involved in implementing the four stages of the Youth Climate Change Adaptation Framework, particularly the

A. Simmons, *Climate Change Adaptation Framework and Youth Entrepreneurship in West Africa*, Sustainable Development Goals Series, https://doi.org/10.1007/978-3-030-85754-7_3

data collection methods utilised during the project's implementation. In effect to commit partner institutions (PI) to the project's operationalisation, the consortium instituted a Memorandum of Understanding (MOU). The processes, pitfalls and successes in the development and implementation of the MOU will also be discussed.

Section 1: The Youth Climate Change Adaptation Framework

The Youth Climate Change Adaptation (YCCA) Framework was adapted for use in SIDS to meet the needs of the young people of Sierra Leone (Simmons, 2021). The framework employed a range of methodologies to collect data about youth in Sierra Leone and the feasibility of utilising the framework to enhance Sierra Leonean youth's employability and entrepreneurship capability. The data collected from the various consultations with key stakeholders helped consultants to develop the three strategic programming areas for urban and rural areas: component 1, skills development for employment (SDE); component 2, career guidance and job readiness (CGJR); and component 3, the Graduate Entrepreneurship Programme (GEP) (Fig. 3.1).

The framework is divided into four steps. These steps include a) identifying development calamities and stressors affecting young people, b) assessing the socioeconomic impact of these development calamities and stressors, c) analysing the skills and knowledge required to enhance youth entrepreneurship and employment and d) adjusting youth development policy, strategy and actions to enhance youth resilience and capacity and promote entrepreneurship and employment (Harley & Minnen, 2010; ILO, 2010; Pereznieto et al., 2011; DFID, 1999 in Angelsen et al., 2011).

Step 1: Climate Change Stressors/ Pressures and Possible Effects

The first step explored the historical and current developmental calamities/pressures impacting youth in the context of Sierra Leone. The theoretical foundation for understanding youth development concerning entrepreneurship and employment issues within Sierra Leone was developed during this stage. Literature reviews provided an understanding of the complexities surrounding how the emerging climate phenomena would impact on LDC like Sierra Leone and offered a range of strategies on entrepreneurship as a means of increasing employment opportunities in the youth sector (Simmons, 2021; UNDP, 2015; Nurse et al., 2014; ILO, 2010).

The consultants collected data from top-down interviews with policymakers and senior officials of government and bottom-up focus group discussions (FGD) with youth, bilateral and multilateral development partners, NGOs and private Sector agencies. Questionnaires were used to gather varying levels of perspectives on climate change and other developmental pressures impacting Sierra Leonean youth (Simmons, 2021). Data were also collected from reviewing reports, policies and other relevant documents.

Step 2: Socioeconomic and Physical Effects of Climate Change on Youth

The second step assessed the socioeconomic impact of climate change and other development calamities/pressures on youth and their communities in Sierra Leone, including the decade-long internal conflict, the Ebola epidemic and the impact of the financial crisis (commodities price crash). This involved investigating the effects of these development calamites/pressures on youth and the communities' socioeconomic situation, health and social life in Sierra Leone (Nurse et al., 2014; IPCC, 2014a, b). The data collected by consultants linked environmental and societal calamities and pressures to youth's overall vulnerability (Government of Sierra Leone, 2019a, b; IMF, 2019; Western Phonebe, 2021; Ripple et al., 2019; Bradshaw et al., 2021).

Step 3: Human Capital Adaptation Process

Step three analysed the skills and knowledge required to enhance human capital adaptation

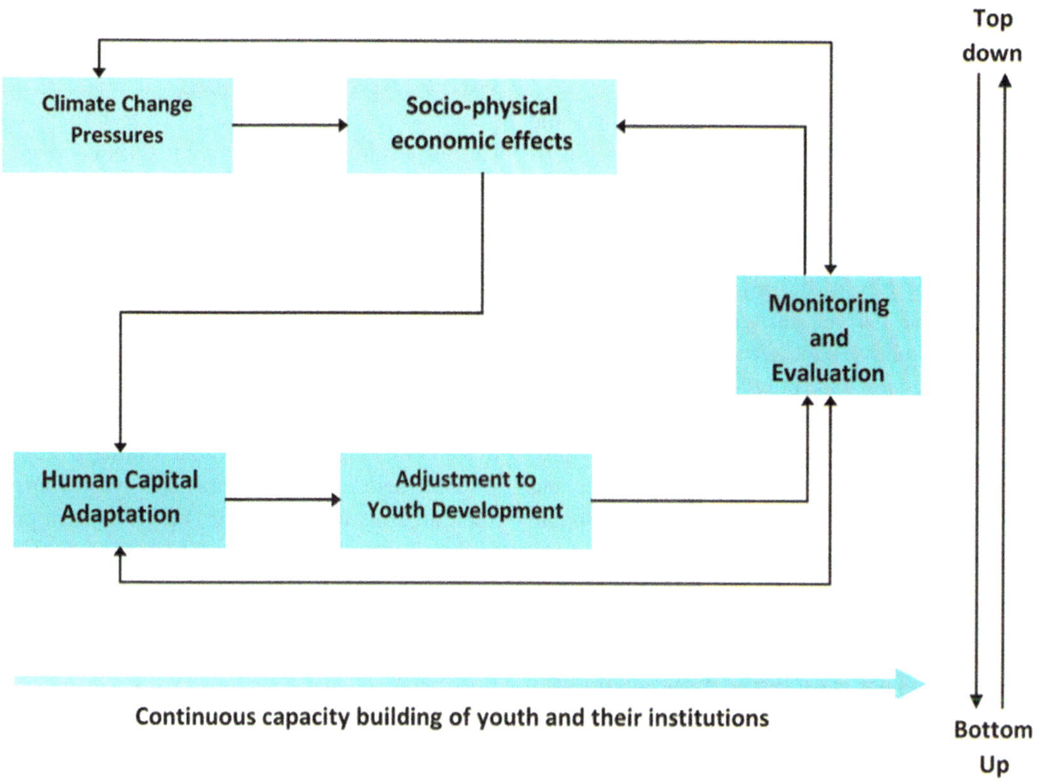

Fig. 3.1 The revised conceptual framework on the adaptation of youth to new climate conditions (Simmons, 2021)

and identify and develop training curricula and modules based on the gaps identified in selected tertiary and TVET institutions across Sierra Leone. The consultants met with officials of tertiary and TVET institutions within Freetown, Bo, Kenema, Makonde and Kinema. The literature review determined the impact of climate change and other development pressures on youth, specifically in entrepreneurship and the job market, which pinpointed a disparity between the market's knowledge and skill needs and those provided through education and training institutions (Margolis et al., 2016). The project consultants then identified and shortlisted education and training initiatives that build youth resilience to adapt to climate change and other development pressures/stressors; three strategic programmes were identified. The TVET curriculum, the methodologies and tools used to deliver employment programmes were also reviewed.

Like the previous stages, data was collected from top-down interviews with policymakers and senior officials and bottom-up focus group discussions (FGD) with youths. Interviews with senior officials from development agencies such as the GIZ, USAID, AfDB and others were conducted. The consultants also reviewed a range of case studies, and best practice approaches implemented in business services centres in Freetown, Kenema and Makeni to gather data on facility functioning. Three institutions in component 1 received new tools and equipment to enhance training delivery and bring it in line with the requirement of the market (Simmons, 2021).

Step 4: Adjustment to Youth Development Policies, Processes and Strategies

Step 4 adjusted youth development policy, best practices and strategies to support the mission of the programme. In this step, consultants worked on developing approaches, strategies and actions to enhance youth capacity and resilience to adapt to climate change and other development calamities.

The consultants conducted reviews on the institutional capacity of the selected partner institutions to implement the project in terms of the development of programmes on youth entrepreneurship and sustainable livelihoods through entrepreneurship in Sierra Leone, including the existing and current curricula, methodologies, practices, processes, tools, management and partnerships. Universities, polytechnics, colleges and TVET institutions worked with the consortium to create mechanisms to adjust and amend processes, policies and strategies to empower and develop youths' resilience to build sustainable livelihoods through entrepreneurship.

The new and emerging processes and mechanisms contributed significantly to enhancing youth and partner institutions' entrepreneurship and employment capabilities. A successful validation workshop was conducted during the inception stage of the project to capture the milestones achieved and the critical issues to be addressed towards completing the pilot phase of the YEEP implementation. A comprehensive monitoring and evaluation (M&E) mechanism was an essential component of the framework. M&E provides robust feedback to policymakers and practitioners to gauge the particular stage's success or health. Its incorporation was crucial in providing feedback to implementers on each stage's overall performance and opportunities to test and modify successive stages (Harley & Minnen, 2010).

Youth Development Outcomes to the Comprehensive Conceptual Framework

This step concluded the overall framework by analysing the literature, reports and case studies on partner institutions' performance within each component and the project's overall outcomes (Appendix 1: The M&E section of the CCA for Youth Development Framework – measuring the performance for the overall project).

Section 2: Memorandum of Understanding (MOU) with Partner Institutions

Partner institutions (PIs) were assessed during the inception stage of the project, and the selected institutions signed MoUs with senior officials of the consortium and NAYCOM. Three institutions were selected for component 1 (Skills Development for Employment). The consortium provided financial resources and technical support to partner agencies to implement training, with responsibility given to PIs regarding facilities and instructors to fulfil said training.

The selected PIs were the Government Technical Institute (GTI) at Ferry Junction, providing training in Sustainable Construction; Njala University Mokonde Campus, providing training in Sustainable Agriculture and Aquaculture; and the Milton Margai College of Education and Technology (MMCT) Brookfields Campus, providing training in Sustainable Hospitalities (Hotel Management and Tourism) for component 1. The Ernest Bai Koroma University in Makeni, Eastern Polytechnic in Kenema, SOS Goderich, Milton Margai College of Education and Technology Goderich Campus and Government Technical Institute at Ferry Junction were selected for the implementation of component 2, providing training in Career Guidance and Job Readiness Skills. The African Foundation for Development (AFFORD) and Global Youth Network and Community Action for Peace and Development (CEPAD) were selected from Makeni, Freetown and Kenema, respectively, to implement activities under component 3, the Graduate Entrepreneurship Programme. The National Council for Technical and other Academic Awards (NCTVA) also signed a MOU with the consortium due to the strategic role it played regarding the development and approval of the new curriculum.

According to the NCTVA website, this agency is a government body which seeks to provide consistent and high-quality accreditation, validation, examination and certification services in education and training to the people of Sierra Leone, through working with institutions, professional bodies, employers and other stakeholders to ensure that the training provided and the qualifications awarded by the Council reflect the needs of the job market. The agency also accredits qualifications awarded by external bodies. Overall, it is envisaged that NCTVA will contribute towards the development of the country's human resource capacity through its services in education and training (https://nctva.org/about-us/).

The MOU is part of the hybrid development approach that combines top-down and bottom-up approaches and provides partner institutions involved in the YEEP with an overall stake in shaping the national development process (Matland, 1995; OECD, 2013; Stochowiak et al., 2016). The PIs demonstrated commitment to youth development in Sierra Leone and view the signing of the MOU as recognition by the government as key leaders in enhancing the national planning and development processes. Due to the MOU increasing confidence, accountability and enthusiasm in the project, PIs were able to increase the numbers of trained instructors in the trades and skills from 20 to 60 instructors.

Partner institutions contributed their limited resources to provide new spaces to train participants in the various skills programmes. The Government Technical Institute (GTI) renovated and made available a separate training room dedicated to the practical course, aluminium fabrication; Milton Margai College of Education and Technology (MMCET) provided a dedicated building for e-travel. The partner institutions have also installed new furniture for the e-learning lab proposed during the TOT workshop, and Njala University's Mokonde Campus launched an association for aquaculture farmers in Sierra Leone during the TOT session.

Component 2 PIs increased the number of trained counsellors in the Career Guidance and Job Readiness skills from 30 counsellors from the Freetown area as stated in the project document to 60 counsellors from 12 partner institutions from districts of Freetown, Mokonde, Bo, Makeni and Kenima. This led to increased participation of young people in the Career Guidance and Job Readiness Skills training from 2400 as indicated in the project document to 6200 young people.

In component 3, the hybrid approach provided partner institutions with the opportunity to contribute to the national development planning process (Matland, 1995; OECD, 2013; Stochowiak et al., 2016). The commitment of PIs to the national development processes led to an increase in instructors from 12 to 18 and participants from 60 to 105.

Reflection

The Climate Change Adaptation for Youth Development Framework was initially developed in the Caribbean SIDS to enhance the capacity and resilience of youth in small island communities to adapt to the impact of climate change (Nurse et al., 2014; Simmons, 2021). The framework was piloted in Sierra Leone during the implementation of the Youth Employment and Entrepreneurship Project (YEEP) in 2018–2020. Within the first month of the inception phase, a comprehensive strategic plan outlining the project, stage activities and deliverables was submitted. The monitoring and evaluation indicators, linked to the steps in the YCCA framework, were also developed to ensure the achievement of the deliverables.

When working in the development sector in Sierra Leone, an important issue for consideration is the level of distrust among key stakeholders. Partner agencies lack trust in officials working in the developmental institutions because of past experiences where officials failed to honour the terms stipulated in contracts. They expressed their unwillingness to participate in the project because on numerous occasions the beneficiaries of projects were not given the resources indicated in the project documents. The officials of the consortium spent a considerable amount of time meeting with heads of institutions, youth and other stakeholders to persuade them to participate in the activities. The consortium agreed with heads of partner institutions to develop a Memorandum of Understanding among the partner institutions as a strategy for committing these institutions to participate in the project.

While the MOU was between a partner agency and the consortium, the Youth Commissioner signed the MOU on behalf of the Government of Sierra Leone. This resulted in boosting the levels of confidence and commitment of partner institutions in the project. Furthermore, there were difficulties on the part of the NAYCOM officials understanding the conceptual framework as it relates to the implementation of the project. The word framework was used loosely by government officials to mean an 'official paper' developed by or for the government on a specific issue or subject. A considerable amount of time was expended by the consortium to ensure that the Government YEEP Project Manager and NAYCOM PIU team understood the concept, its application within the overall YEEP project management cycle, the steps in operationalising each of the stages of the framework within the project management cycle, the relationship and linkages between the stages of the framework and the stages and deliverable of the project.

During the project's implementation period, there were moments of considerable delays in the funding to support the implementation of the project activities in the field. This was because of a) delay in the approval of the stages of the project by NAYCOM and b) the consortium delay dispersal of funding to the implementation team in Freetown as a result of the delay in the approval of the project state. Some of the officials of the partner institutions chose not to implement the project activities until the funding arrived even though the MOU committed the consortium to provide the financial resources required to implement the training programmes. The consortium took a considerable amount of time to visit each partner institution to persuade officials to utilise their own resources, which would be refunded when the payment is released. Over three-quarters of the partner institutions did not have the financial resources to fund the implementation of these courses.

The delays in funding of the project created considerable setbacks in the implementation of component 2 of the YEEP. There was one instance mid-way through the project where only two of the seven institutions that were delivering the component 2: Career Guidance and Job Readiness Skills training course was actually delivering the course. After 3 months of physically driving to these campuses to negotiate with officials of partner institutions to resume the training, four out of the five remaining partner institutions finally agreed to kickstart their training activities. The fifth lost interest and all future communications were terminated. The NAYCOM project manager and the monitoring and evaluation managers also visited partner institutions to investigate whether the project's deliverables were delivered adequately. The lack of communication and coordination of this activities resulted in tensions between the officials of the consortium and NAYCOM.

In March 2020 due to the COVID-19 pandemics, the country went into lockdown. This significantly hindered the project's progress as the programme had to be reviewed and amended to suit distance learning. Universities, colleges and TVET centres were on closed from March until July 2020. In May 2020, the consortium persuaded the PIs to resume their project even though these institutions were officially closed. Partner institutions were encouraged to deliver their training courses through e-learning, although there were serious challenges in the implementation of training due to poor Internet connectivity. Institutions did not have the technology to deliver the training, and over 70% of students had no access to smartphones, tablets nor computers to access training. Institutions, however, were able to engage with students by being highly creative in setting up study groups and using WhatsApp and Facebook apps and platforms to deliver training resulting in some institutions doubling and tripling the number of youths participating (Thuraya, 2015; UNDP, 2015; Simmons, 2021). The MOU document is located at the end of the chapter (Appendix 1).

Section 3: The Public-Private Partnership and YEEP Pilot Model Conceptual Frameworks

The Public-Private Partnership Framework

The Government of Sierra Leone recognised that skills development is essential to youth employment and job creation in Sierra Leone (Government of Sierra Leone, 2004, 2015, 2019a, b; World Bank, 2017; IMF, 2019). The consortium embarked on a comprehensive study, followed by a series of consultations on the youth employment challenge in Sierra Leone, culminating with a process of defining the roles and strategies to be implemented to ensure PPP in TVET. The consortium employed multiple research approaches to collect data from key stakeholders of government agencies, private sector companies and TVET institutions on the policies, processes and mechanism to operationalise PPP in the TVET sector (Chan et al., 2005, 2010; Fengyu et al., 2020). The consortium conducted desk reviews by analysing and synthesising key reports and documents, including labour and household surveys, education reports and economic reports (check Chap. 4 research for details on the data collection methods used to collected data).

Key informant interviews were also conducted with private sector officials, TVET institutions and other partners to assess partnership processes and relationship; identify what worked, policy and strategic options available; and make recommendation where possible to enhance policies, mobilisation of human and financial resources and other possible future changes. FGDs were conducted with youth to solicit their view on the government and private sector agencies' collaborative roles to enhance the quality of TVET programmes in the future. Youth also provided their views on the strategies, relevance and quality of the courses provided by PIs and what can be done to add value. The study discovered that youth in Sierra Leone have inappropriate skills to enhance their employability due to the mismatch between institutions' training and those required by the market (Margolis et al., 2016, Carter, 2008). This issue was compounded by the non-availability of appropriate tools, irrelevant curricula and lack of appropriate trained instructors and counsellors to effectively deliver the training to youth in partner institutions (ILO, 2010; Poschen, 2015).

The Government of Sierra Leone considers the PPP as appropriate because it can reduce the country's skills gap and promote youth employability (Government of Sierra Leone, 2015, 2019a, b; IMF, 2019). It is also linked to the Government of Sierra Leone National Development Plan (Government of Sierra Leone, 2019a, b). Therefore, to strengthen the implementation approach, the consortium established partnership with the TVET secretariat to develop the PPP framework and encouraged private sector investment and participation in the development initiative. The PPP framework utilised the hybrid approach (Matland, 1995; OECD, 2013) developed as a guide reflecting the principles, processes and strategies to strengthen the relationship and partnership between the public, civil society organisations and the private sectors to collaborate as partners in the development, implementation and delivery of TVEP initiatives (Poschen, 2015; ILO, 2010).

Reflection

The PPP framework was developed to guide all stakeholders participating in the development and implementation of the TVET process in Sierra Leone. During the initial stages of the consultation with the private sector and civil societies, the agencies were sceptical to participate in any discussion on the development of the PPP policy document because they believed that developing and providing technical vocational education training programmes are the responsibility of the government. During the meeting with private sector agencies, there were numerous complaints regarding the poor quality of young professionals produced by TVET institutions in Sierra Leone. The senior officials of businesses highlighted the mismatch between the skills that they require in the workplace and those that are produced by the education and training institutions (Carter, 2008; Poschen, 2015). These agencies became incredibly supportive once convinced by the consortium staff of the benefits to be derived from this partnership, including the opportunity to add value and shape the curricula regarding students' skills and knowledge to produce work-capable individuals. The idea to develop a MOU to commit the private sector and government to support the YEEP initiative evolved from the PPP consultative process.

The private sector was invited by the consortium to participate in the development of the curricula for training youth in skills for employment initiatives for components 1, 2 and 3, including Skills for Employment, Career Guidance and Job Readiness and the Graduate Entrepreneurship Programme. This was sanctioned by NAYCOM due to the link in the National Development Plan (Government of Sierra Leone, 2019a, b; IMF, 2019). A training manual was developed in collaboration with PIs delivering component 2 to facilitate training to more than 6200 students in partner institutions. The training manuals for all the other training courses were developed with input from the private sector and approved by the NCTVA.

An online e-learning platform was established as a swift response to the COVID-19 restrictions and to ensure the continuation of e-learning activities on a virtual platform (Thuraya, 2015). The cost for delivering the new virtual courses was supported by PIs as part of their contribution to the national development planning process. Shifting of training delivery from direct classroom training to e-learning provided opportunities for the training to be delivered to more youth than was initially anticipated and allowed greater meaningful interaction between lecturers and students. Although it is still early to assess the impact of the PPP framework on the quality of TVET education in Sierra Leone, there is a high level of enthusiasm on the part of the private sector to work with the government to provide resources such as tools, training materials and internships and apprenticeship placement opportunities to support TVET development in Sierra Leone. Therefore, the government should build on the opportunity provided by the YEEP to engage the private sector's participation in the delivery of TVET to youth across Sierra Leone. A detailed summary document of the PPP framework is in Appendix 2 of the end of the chapter.

YEEP Pilot Model Conceptual Framework

The Government of Sierra Leone recognises the importance of the YEEP as the vehicle to address the severe unemployment challenges facing youth and to enhance their contributions towards achieving the national sustainable development goals (Government of Sierra Leone, 2019a, b). The YEEP pilot model framework is a conceptual 'blueprint' document highlighting the processes, structures and mechanisms required by the stakeholders to implement youth employment and entrepreneurship development programmes involving a hybrid approach, i.e. a top-down and bottom-up approach to resolve the unemployment problem through entrepreneurship (OECD, 2013, Stochowiak et al., 2016).

Reflection

Climate change is the most critical development challenge faced by the planet today (Nurse et al., 2014; IPCC 2014a; Jaja et al., 2016; Western Phonebe, 2021; UNEP, 2021). Climate change affects all areas of life and livelihoods on earth. Scientist warns of a ghostly future of mass extinction and climate disruptions (Western Phonebe, 2021; UNEP, 2021).

The Government of Sierra Leone, through the NAYCOM, is in the process of implementing the upscaling of the YEEP and working with the consortium to document the processes and strategies utilised in the implementation of the project through the pilot model framework. The pilot model framework presents the knots and bolts of implementing the YEEP. It describes the mechanism for operationalising the YEEP and related youth employment projects. The framework employed a range of methodologies to support data collection and assess the performance of the various stages of the climate change adaptation for youth entrepreneurship and the deliverables and stages of the YEEP (Simmons, 2021). The data collected from consultations, interviews, questionnaires and FGDs with key stakeholders assisted the formulation of YEEP, the strategic programming areas, timelines and milestones.

Due to the successful implementation of the YEEP project, the African Development Bank (AfDB) committed itself to funding the upscaling of the YEEP into a national programme to enhance the employability of Sierra Leonean youth. The government will use the YEEP pilot model framework, which evolved from the pilot, to implement the national project. A detailed summary document of the pilot model framework is in Appendix 3 at the end of the chapter.

Summary of the Chapter

The Climate Change Adaptation Framework Youth Entrepreneurship (CCAFYE) operationalisation was deliberated in this chapter. Each stage of the YCCAFYE, including issues, challenges and strategies and the data collection methods, was discussed (Simmons, 2021). At YEEP's initialisation, there were some difficulties getting NAYCOM PIU to accept the YCCAFYE as the conceptual framework for implementing the project. The confusion was caused due to difference in the comprehension and appreciation of the framework by the NAYCOM officials who were responsible for the management of the project on behalf of the government of Sierra Leone. This resulted in considerable time lapses and delays in approval and implementing the first and second stages of the project. This confusion was resolved by the direct intervention of the Youth Commissioner.

The consortium utilised a series of consultative strategies to engage stakeholders in the project including field visits and district and national stakeholders' consultation workshops which led to the signing of MOUs with partner institutions. MOUs with PIs participating in the project's implementation strengthened the partner agencies' commitment to the project. The consortium provided financial and technical assistance and training to partner agencies to support implementing training programmes in their institutions. NAYCOM provided tools to partner institutions involved in component 1 –Skills for Employment training – as well as ensuring that the indicators for the M&E were achieved. The chapter discussed the actions taken by the staff when the consortium could not provide the financial resources on time to partner institutions to implement activities due to delays in approval of the stages of the project. Some of the PIs agreed to continue their training until the resources became available, while others decided to halt the training until the funding was provided.

To alleviate some of the financial shortages in implementing the vocational educational training programmes, the consortium's PPP framework encourages partnerships between interested private sector organisations and Sierra Leone's government to mutually enhance human capacity and financial resources. The government gained commitment of more funding to increase the number of youths involved in education and human capacity development activity, while the private sector gained an increased professional candidate pool, minimising the current skills gap, as they directly shaped the training to suit the market's needs. Both sectors benefited mutually be an improved society's capacity and quality of youth workers. The YEEP pilot model framework supports the PPP framework, which provides an overview of the stages, procedures and processes under consideration when establishing YEEP or similar youth development and sustainable livelihood projects. These policy framework documents in tandem are essential tools to use for implementing such projects.

Appendices

Appendix 1: Memorandum of Understanding

The Undersigned

(Implementing agency) (include the names of the members of Consortium) (hereinafter **(Implementing agency)** is an international consultancy firm whose administrative offices are at **Address** and organised and existing under the laws of country, herein legally represented by its Directors. State brief information about the names of the consortium.

[**Name**] (hereinafter **Development Partner Institution**"), a [legal entity] organised and existing under Sierra Leonean law and with its principal place of operations located at address, herein legally represented by its duly authorised official.

(**Name of the agency of the Government of Sierra Leone**) (hereinafter "**Contracting Authority**"), a [legal entity – **Government of Sierra Leone**] organised and existing under Sierra Leonean law and with its principal place of operations located at address, herein legally represented by its duly authorised official.

Each individually referred to as "Party" and collectively as the "Parties";

Recitels

Whereas (**Implementing agency**) has been awarded a contract by the (**Name of the agency of the Government of Sierra Leone)** to support in the implementation of the Sierra Leone Youth Entrepreneurship and Employment Project (Sierra Leone YEEP) to provide support to the (**Name of the agency of the Government of Sierra Leone**) to implement this project, starting 1 of July 2018 ("**Project**").

Whereas (Name of the agency of the Government of Sierra Leone) has contracted (Implementing agency) to support (Name of the agency of the Government of Sierra Leone) to implement the Sierra Leone Youth Entrepreneurship and Employment Project (Sierra Leone YEEP).

Whereas name of TVET Institute **(herein after referred as a training provider) has been identified (Implementing agency) and approved by (Name of the agency of the Government of Sierra Leone) to provide implementation support to execute Component 3 of the Project (Graduate Entrepreneurship Programme).**

Whereas the Project aims to reinforce the employability of Sierra Leone Youth as well as enhancing their entrepreneurship skills through the development of a framework which integrates education and vocational institutions with job market demand and supply. It is envisaged that this approach will result in strengthening the occupational levels in the youth population segments of the country which will eventually trigger sustained economic development in the growth sectors. The project has three main components as depicted in below figure:

Whereas the component is one of the three main activity lines of the project. Whereby Component 1 aims at building capacity for training youth in three specific priority skills (as will be determined in consultation with TVET coalition[1] partners) needed by private sector employers and other operators in growing sectors of the economy and linking them to the private sector for employment. (**the Pilot**).

Whereas the Project is led by a project team from the (**Implementing agency**) in close cooperation with (**Name of the agency of the Government of Sierra Leone**). ("**Project Team**").

Whereas the Project team seeks to partner with existing Business Support Centre (BSCs) for the Concept Development with the perspective of implementing the Graduate Entrepreneurship Programme with these partners (include collaborating training institutes).

Whereas (**Implementing agency**) and **[Name TVET Institute]** have been discussing the Project and share their vision on the Concept.

Whereas [Name TVET institute] has proven to fit the profile of a Business Support Centre with focus on support youth entrepreneurs to improve their business.

Now, therefore, the Parties wish to further explore areas of collaboration in accordance with the terms of this Memorandum of Understanding ("**MoU**") on this [date] (the "**Effective Date**").

Purpose of this Memorandum of Understanding

To record our mutual understanding of cooperation in the delivery of services.

Principles Guiding the Development Partnership

The objectives of the component are:

[1]Public-private sector partners in TVET sector

Conducting a Training of Trainers for 20 TVET instructors for the priority skills and trades and entrepreneurship training, including developing a TVET instructor apprenticeship/ internship programme in cooperation with National Council for Technical Vocations and other Academic Awards (NCTVA).

Training a minimum of 100 youths in supervisory skills and such youth should have work experience and have potential to excel.

Training 150 youth in the selected priority skills. *This will include going through a graduate internship programme which will be conducted in cooperation with UNDP and the National Youth Commission and will also include linking the trained youth with private sector employers for employment.*

Conduct capacity building of selected TVET institutions in management and governance.

The expected results are:

- Reviewed and updated course content and training methodologies for the three priority skills and trades in selected TVET institutions.
- Developed entrepreneurship curriculum for the selected TVETs.
- Training of Trainers for 20 TVET instructors for the priority skills and trades and entrepreneurship training.
- Training of a minimum of 100 youths in supervisory skills.
- Training of 150 youth in the selected priority skills; and.
- Capacity building of selected TVET institutions in management and governance.

Scope

- To design, develop and implement three priority sector skills training for youths to and from both parties consistent with this MOU.
- To identify and deliver appropriate entrepreneurship and skills training, monitoring and mentoring for entrepreneurs and enterprises.

- Organising of training programmes in various skills trades taking into account the need of labour market in identified sectors.
- Where possible to standardise the delivery of services to youths and entrepreneurs.
- The institution shall play a key role in skills development in three key trades, skills upgradation, innovation and competitiveness of youths applicable of an industry. Both parties believe that close co-operation between the two would be of major benefit to the youth learners to enhance their skills and knowledge.
- **Curriculum Design: The partner institution** will give valuable inputs in teaching/training methodology and suitably customise the curriculum so that the students fit into the industrial scenario meaningfully.
- **Skill Development Programmes: The partner institution** is to train the students on the emerging skills identified in order to bridge the skill gap and make them industry ready.
- **Teacher Development: The partner institution** is to train the teachers for imparting training as per the industrial requirement considering the NCTVA in concerned sector, if available.
- **Placement of Trained Students: The partner institution** will actively engage to help the delivery of the training and placement of students into internships/jobs; and will facilitate placements for at least ......% of the students.

Intellectual Property Rights

Each Party will retain title of all intellectual property rights it already holds under applicable laws, and nothing herein will be deemed to be a transfer thereof or a licence thereupon.

The intellectual property rights regarding tools, formats, knowhow or any other developed documents, generated by one of the Parties specifically for the execution of this partnership, remains with that particular Party, unless specifically agreed otherwise.

De Property Rights of the framework will be owned by all parties.

Financial Arrangement

Payment milestone	Timeline	% of contract value to be paid out on achievement	Contract value in USD
Upon submission and approval of selected list of trainers for the training, CVs of trainers, training plan	June 2019	10%	
Upon approval of report on completion of the training, detailed work plan and schedule recruitment, assessment and training of participants, curriculum readiness, training methodology, CBT training materials development plan, etc.	July 2019	10%	
Upon approval of report on completion of selection of youths for priority skills training and training readiness plan (workshop, tools and equipment, trainers, materials, etc.)	July 2019	45%	
Upon approval of report on completion of training of 150 youth and placement of trained youths.	August 2019	20%	
Upon approval of report on submission of training evaluation report	August 2019	15%	
Total		100%	US$ 10000.00

Confidentiality

Each Party (the "Receiving Party") agrees that, unless otherwise stated, any information or materials provided by the other Party (the "Disclosing Party") are confidential and proprietary to the Disclosing Party and will be kept confidential by the Receiving Party. Such information shall not be used by the Receiving Party for any reason beyond the purpose of this MoU. The Receiving Party will have no duty of confidentiality with respect to information that (i) was rightfully known to the Receiving Party previously, without a duty of confidentiality; (ii) is, or subsequently becomes, legally and publicly available without breach of this Agreement; (iii) is rightfully obtained by the Receiving Party from a third party without a duty of confidentiality; or (iv) is disclosed by the Receiving Party under a valid order created by a court or government agency, provided that the Receiving Party provides the Disclosing Party with prior written notice of such obligation and the opportunity to oppose such disclosure.

The terms and conditions of this MOU shall also be considered confidential and proprietary information that shall not be disclosed by either Party unless mutually agreed.

Media and Communication

Parties will align their communication and publicity strategy with reference to the MoU. Media statements, promotional material and other publicity referring to the other Party and/or the MoU may only be published after obtaining written consent of the other Party.

Duration and Termination

This MoU is entered into force from the effective date to May 2019, unless terminated in writing with a notice period of thirty (30) calendar days. This term aligns with the timeline for implementation of component 1 (30 September 2019) and can be extended by signed addendum should the timeline be delayed.

Any Party may terminate this MoU immediately upon notice to the other Parties if one or both of the other Parties commit an act that in any way denigrates the reputation of the terminating Party or dilutes the value of the terminating Party's brand, trademarks or logos as determined in the terminating Party's sole discretion.

Changes or additions to this MOU, including the appendices, shall be valid only if agreed upon in writing by the Parties.

Reference Law

Any matter relating to the interpretation or application of the MOU which is not covered by its terms shall be resolved by reference to the UNIDROIT Principles of International Commercial Contracts (2004).

Language

This MoU is drawn up in the English language. If the MoU is translated into another language, the English language text shall in any event prevail.

Arbitration

Any and all disputes arising from or in connection with this MOU, or further contracts resulting there from, shall be finally settled in accordance with the Rules of Arbitration of the International Chamber of Commerce by one or more arbitrators appointed in accordance with the said rules. The place of arbitration shall be City. The arbitral procedure shall be conducted in the English language.

Invalidity

If at any time a provision of this MOU is or becomes illegal, invalid or unenforceable in any respect under the law of any jurisdiction that shall not affect or impair:

The legality, validity or enforceability in that jurisdiction of any other provision of this MOU; or:

The legality, validity or enforceability under the law of any jurisdiction of that of any other provision of this MOU, and.

The Parties shall replace the non-binding provision with another provision that is binding, in such a way that the new provision differs as little as possible from the non-binding provision, taking into account the object, the intent and the purpose of this MOU.

Entire Agreement
This MOU contains the entire agreement between the parties in relation to the matters contemplated by this MOU and any such document and supersedes any previous agreements between the parties in relation to such matters. This MOU can be amended, restated or supplemented only by an instrument in writing signed by all Parties.

Each of the Parties confirms that in entering into this MOU, it has not relied on any statement, representation, warranty, agreement or undertaking of any person (whether a party to this MOU or not) other than those expressly set out in this MOU and that it will not have any claim, right or remedy arising out of any such statement, representation, warranty, agreement or undertaking.

Miscellaneous
Nothing in this Memorandum of Understanding will create a relationship of partnership, agency or joint venture between the Parties. Neither Party is authorised to act or make any statement, representation or warranty on behalf of the other Party.

Each Party will comply with all laws, regulations, rules and ordinances applicable to it in connection with the transactions governed by this MoU. The provisions of this Memorandum of Understanding that, by their nature, are intended to survive any expiration or termination of this MoU will survive such expiration or termination. A notice under this MOU shall only be effective if it is made by registered letter.

Unless expressly agreed otherwise, all correspondence related to this MOU shall be worded in English and shall be addressed to the above-mentioned representatives. Each of the parties is obliged to notify the other Parties of a change of address at any time. Each party chooses as its address for all purposes under this MOU, or for serving of any court process or documents, giving any notice, or making any other complication of whatever nature and for any purpose as follows:

(Names of Implementing organisation)
address
Tel:
[name]
[Address].

The following persons shall be responsible for liaising over the implementation of this MOU:

- A representative of (Implementing organisation).
- Representatives of **TVET Institute**: [name], [function].
- A representative of (Government Institution).

In witness whereof, each of the Parties hereto has caused this Agreement to be duly executed and delivered by a duly authorised officer as of the date first above written.

(Implementing organisation)	
Name:	
Function:	
______________________	
Date:	
(name of the agency of the government of Sierra Leone)	
Name:	
Function:	
______________________	
Date:	
(name of partner institution)	
Name:	
Function:	
______________________	
Date:	

Appendix 2: Summary of the PPP Framework

Introduction

Background
Getting youth into productive employment is an urgent policy issue for countries around the world. A successful transition to the labour force is essential for young people to be assured of success in life; however, many young people encounter significant obstacles in this transition.

Governments in low- and middle-income countries, such as Sierra Leone, are actively engaged in policies to help youth attain the skills they need to acquire decent jobs and sustainable livelihoods (ILO, 2010). The range of policy options to achieve the above is wide and includes skills training, employment intermediation services to place workers with firms, self-employment promotion and provision of subsidies to firms to hire youth, among others.

Increasingly, governments are recognising that the private sector and employers have a vital role to play in developing and implementing youth employment and sustainable livelihood programmes which lead to the employment of youth. Partnerships with the private sector would inject financial and technical resources beyond those that governments can mobilise to finance such programmes. The private sector can provide services to bring curriculum and training in line to what is required within the labour market. It is important to note that the active engagement of employers in the development and provision of skills in the education system (formal and informal) would enable alignment with the requirements in the labour market. In essence, the technical vocational education and training (TVET) systems in the least developing countries such as Sierra Leone were deemed ineffective and irrelevant to the development needs of the market. To some extent, this is due to the absence of a deliberate policy that enables the private sector to participate in the design and development of TVET curricula and the delivery of training content and standards.

The impact of climate change and other development calamities is excluded from the design and development of TVET and future occupations/livelihood jobs in most developing countries in Africa (ILO, 2010). Therefore, this needs to be an essential consideration when developing TVET and other forms of youth skills training and apprenticeship programmes in the future (UNDP, 2015). PPP in the TVET sector can take several forms when piloting new initiatives, training partnership, sectoral partnership and system-wide partnership. Therefore, the purpose of the PPP framework for skills development is to explore the policy, strategies and process and the benefits to stakeholders for the government of Sierra Leone to work in partnership with the private sector in the development and delivery of the TVET sector.

Definitions and Concepts

TVET is concerned with the acquisition of knowledge and skills for the world of work. In the past, various terms were used to describe elements of the field that is now conceived as comprising TVET. One definition is that TVET is any education, training and learning activity leading to the acquisition of knowledge, understanding and skills relevant to employment or self-employment (Poschen, 2015; UNDP, 2015). TVET serves here as an overarching term to describe all kinds of formal, non-formal and informal training and learning provided by or in all different institutions, providers and learning locations.

Private Sector

Over the past decades, the private sector has been involved in multiple ways to promote youth employment and skills, including but not limited to:

- Partnering in initiatives to train youth by providing funding, developing course content, contributing to teaching and providing on-the-job experience to trainees.
- Funding entrepreneurship promotion programmes and supplying credit, grants and technical assistance to young entrepreneurs.
- Engaging in high-level planning for training and employment strategies with government and other stakeholders.
- Supplying training services under competitive contracting with the public sector or with employers.
- Developing inclusive value chains in agriculture and other sectors involving young entrepreneurs.

The involvement of the private sector in youth skills development and employment is a complex issue, not just because the modalities of this involvement are diverse but because the nature of the firms and their motivations vary significantly. Therefore, comprehensive analysis is vital to design an appropriate framework relevant to the national development context.

Private Provision
Public-Private Partnership (PPP) has been employed extensively for the development of infrastructure projects around the world due to its effectiveness in delivering value-for-money (VFM) (Chan et al., 2010). The education and training markets are characterised by a diversity of providers across different subsectors, ranging from for-profit commercial enterprises, religious entities and non-profits run by nongovernment organisations (civil society); publicly funded establishments such as government-aided schools that are operated by private boards; and community-owned operations. UNESCO regards as "private" any educational institute that is controlled and managed by an NGO, i.e., including religious group, association, enterprise or whose governing body consists mainly of members not selected by a public agency (ILO, 2010).

Private Sector Partnership (PPP)
Public-Private Partnership (PPP) is generally defined as a system in which a government service or private business venture is funded and operated through a partnership to deliver a project or service traditionally provided by the public sector. PPP is a business model which aims at financing, designing, implementing and operating TVET through innovative methods (Kruss et al., 2017).

PPP in TVET
Public-Private Partnerships have been defined in a wide range of ways in the policy and academic literature. PPPs are defined as an agreement between the government and one or more private partners, which may include the provision of finance and technical assistance to run the entity. This may concern the private partners delivering the service in such a manner that the service delivery objectives of the government are aligned with the profit objectives of the private partners (OECD, 2008 in Kruss et al., 2017). PPPs are also perceived as a strategy to achieve a social agenda, such as solving a social problem in a community. It is a strategy that aims at bringing the voices that have been on the periphery into the national development planning process while achieving efficiency in planning, management and monitoring.

Public-Private Partnerships (PPPs) in TVET have been receiving widespread attention over the past decade as a system to enhance the future sustainability of the TVET sector. The competing demands on the state's limited resource compounded by an increasing demand upon the provision of educational services are tremendous pressures on the government to explore innovative opportunities to address the gap in the financial resources to sustain the education provision. The private sector engagement in the TVET provides an excellent opportunity for the private sector to share a more significant role in the financing the provision of education (Kruss et al., 2017).

Overview of TVET in Sierra Leone

TVET in Sierra Leone

Sierra Leone is categorised as a small, low-income and fragile country located in West Africa with a population of 7.4 million and gross domestic product (GDP) per capita of US$496 (2016). Sierra Leone's population is young and growing rapidly, and a majority live below the poverty line. Between 2004 and 2015, the population grew an average of 3.3% per annum. Almost one-half (45.8%) of the population is under the age of 15. In 2014, the labour force participation rate was 65%, with only 2.8% unemployment (World Bank, 2017; Government of Sierra Leone, 2015).

TVET in Sierra Leone has the potential to enhance youth employability and improve economic growth by addressing gaps in the labour market. In relation to the higher education sector, TVET focuses on providing practical training, knowledge and skills development directly linked to economic development, specifically occupations, production, services and livelihoods.

The Government of Sierra Leone has long recognised the importance of TVET. NCVTA, as the national autonomous apex body for the TVET sector, is responsible for skill development policy implementation. The NCTVA lacks the human and technical capacity to implement an effective TVET system. The TVET system in Sierra Leone also encounters numerous development challenges, including a) several actors in the system are not functioning, resulting in fragmentation and coordination challenges; b) private sector employers are not substantially involved in developing the TVET system, and as a result, TVET supply is disconnected with the demands of the private sector and the market; and c) the combination of a fragmented TVET system with limited employers and sector engagement means that the overall quality of the training and products delivered through the TVET sector is inadequate. There is currently no formal mechanism to assess the actual demands for skills in key economic sectors.

In its effort to bridge the mismatch in the skills produced by the education and training institutions and the skills required in the workplace, the government mobilised funding and technical resources from World Bank, GIZ, AfDB and other multi- and bilateral agencies to institute initiatives to enhance the overall skills development programmes and institutions across the country. The PPP framework evolved as a critical area of investment by the government in the human capital of Sierra Leone (Table 3.1).

TVET and Economic Growth and Development

There is a conscious effort on the part of Sierra Leone's government to diversify its economic base by investing in the creation of the enabling environments in the sectors that exhibit potential for growth and economic development. The government lists these sectors as priority growth areas, namely, agriculture, tourism, construction, small and medium enterprises (SMEs) and mining. The New Direction policy document - Vision 2025 - and the Agenda for Prosperity include the areas the government will target for driving economic development and the creation of gainful employment for youth. In keeping with its objective of sustainable socioeconomic development, the government is dedicated to supporting these critical economic drivers.

Despite the concerted efforts on the part of the Government of Sierra Leone to hasten the economic growth process by enhancing employment opportunities, youth unemployment continues to pose fundamental challenges. Unemployment among the educated youth who have graduated from universities and colleges is also becoming a serious concern of officials. It is believed that the persistent unemployment among graduates in Sierra Leone is due mainly to the mismatch between the skills and knowledge provided by tertiary institutions and that which are required by the job market (Carter, 2008; UNDP, 2015).

Research shows that there is consensus on the relevance of TVET systems to develop the skills that are required by the workforce and to consequently boost productivity, economic growth and high-quality employment (ILO, 2010, 2013; UNDP, 2015). Research shows a close relationship between the provision of quality TVET, particularly effective practical training, apprenticeship and internship systems, to producing viable skilled youth to address the gaps in the labour market (Poschen, 2015; Simmons, 2021).

TVET and Access, Quality and Relevance of Skills

The quality of skills development programmes provided to youth in Sierra Leone is extremely limited in terms of programme content, methodology and the lack of modern training tools at the training institutions delivering TVET initiative (Carter, 2008, UNDP, 2015). This is compounded by the low capacity in the country to absorb the increasing numbers of unemployed youths and school dropouts (Government of Sierra Leone, 2015).

The skills development programmes provided to youth do not meet the aspiration of the youth and the need of the marketplace. Many young people do not have the soft skills required in terms of time management, job application preparation and participation in job interviews to

Table 3.1 The issues and challenges of implementing TVET

Issue	Components
Relevance and quality: Lack of economic relevance with insufficient skills produced	Insufficient industry participation in policy, planning, needs identification, etc. Lack of labour market information about demand and trainee employment rates Rigid supply response – Lengthy programmes, improper targeting Inadequate attention to skills development for the informal sector Insufficient enterprise training, lack of awareness and inadequate use of incentives Inadequate use of technology-based materials to demonstrate simulations of actual workplace
Access and equity: Inequitable access to skills provision	Technical and vocational education and training focusing mainly on unemployed youths, adults, out-of-school and the disadvantaged Relatively low percentage have access to technical and vocational education and training Females are under-enrolled, and gender bias exists in public provision Barriers to access for low-income households, e.g. unaffordable fees or minimum grade pass Few programmes to raise skills and income of those in the informal economy
Teaching quality: Skills not properly taught – Ineffective skills acquisition	Shortage of qualified teachers and instructors Lack of opportunities for in-service training to upgrade qualifications Predominance of time-based versus competency-based programmes Practical skills not well provisioned Examinations emphasise theory over practical skills Funding mainly for salaries; little for equipment, consumables Lack of practical facilities, equipment and consumables Inadequate involvement of industry-experienced trainers and limited opportunities for continuous upgrading of practical skills
Governance and management: Ineffective organisation and management	Plans and policies mutually inconsistent and lacked costing for feasibility Lack of authority to act by public institutions Weak quality screening and monitoring of private providers Weak information base for research, policy and monitoring
Mobilisation of resources: Inadequate financing for skills and inefficient use of funds	Overreliance on public financing Low levels of public financing Lack of enterprise contributions and income generation at training institutions
TVET image	Students have a poor perception of skills training and view it as an option of last resort Students and their parents lack funding or unwilling to pay for skills training

enable them to maintain a job. This has led to some programmes being undersubscribed and TVET graduates not taking up related employment. This situation has exacerbated the poor social acceptance of TVET, which has been affected by several factors such as poor working conditions and low wages of TVET occupations.

Ineffective Training Delivery

Employers have criticised training delivery of the TVET institutions as not being effective, as they do not prepare jobseekers with the appropriate skills and attitudes needed for employment (Carter, 2008). Employers expect higher competency in core skills as well as soft skills of the TVET graduates. This can be attributed to weak linkages between the industry and the institutions on the job and the absence of setting facilities and equipment standards for implementing TVET.

Ineffective Management and Administration of Training Delivery

The revised Education Act (Government of Sierra Leone, 2004) has designated the National Council for Technical and Vocational Education as the coordinating body for TVET but has not been established. As a result, the monitoring of TVET is still weak for quality and relevance.

The Financing of TVET

Although TVET is recognised as the backbone of a country's economy and strategy for employment generation, there is weak and ad hoc advocacy for TVET by top-level leadership, i.e. at the national level. There is no widespread awareness of TVET reforms and initiatives, which has led to insufficient funding to support the development and expansion of the TVET subsector over the past decade. Currently, the government is putting mechanisms in place by mobilising technical and financial resources from World Bank, GIZ and African Development Bank to ensure equity and sustainability in TVET funding.

Role of Private Sector in TVET

Integrating labour market needs into the TVET system and linking it with better outcomes is a pressing issue in TVET provision today in Sierra Leone. Private stakeholders are rarely involved in shaping framework conditions for skills development and technical vocational education and training. This calls for elaborate policy on the opportunities and challenges for public authorities to create favourable framework conditions for private-sector engagement in TVET delivery, explicitly looking at opportunities to institutionalise the cooperation between private and public stakeholders in integrating learning and working at the workplace.

Therefore, to improve the responsiveness and relevance of training, skills training strategies need to be implemented to pursue public-private partnership in TVET, to expand work-based and enterprise-based training and to strengthen the links between public and private employers when hiring TVET graduates (UNDP, 2015; ILO, 2010). Currently, there is no tangible policy in Sierra Leone to provide incentives and rewards to generate broader industry support and commitment for training, expanding scholarships and other assistance to fund the development of critical skills and higher technologies and incentivise the TVET institutions. Critically, it is necessary to increase the prominence of private sector agencies in shaping occupational standards and the government's policy on TVET.

There is a tendency globally to focus on skills development within the private sector; however, there are many issues that the private sector in Sierra Leone do not possess the capacity to respond to effectively. These include developing national policies, focusing priorities for economic development, guiding curriculum development, quality assurance systems and updating national labour market information systems. These areas are the government's responsibility; therefore, this is a unique role they can play in the new PPP framework.

Public-Private Partnership (PPP) aims to change the role of the public sector in TVET systems and requires a conceptual shift from implementers to facilitators of change. This changing government role includes shifting from the provider of TVET services and programmes to the role of facilitation and regulation of TVET. The TVET PPP process provides increasing opportunities to the private sector to take responsibility for skills provision and standards required for human resources development to support national and regional economic development.

Mapping of PPP in TVET

Current Situation of Private Sector Engagement in Skills Development.

The TVET sector in Sierra Leone is weak, complex, with a wide range of public and private actors. While there have been efforts to create TVET standards and improve coordination, many of them do not sufficiently address the sector's needs. Some of the fundamental building blocks for a strong TVET sector are missing; much of this is related to a lack of national qualification frameworks, definitions and labour market information. There is a need to address the issues of developing occupational standards defined by the private sector and a system of occupational classification distinct from training/curricula classification. There is confusion in Sierra Leone in terms of defining and applying the concept of occupational classification and standards in TVET. Such basic definitions need to be clarified to improve the fundamental building blocks regarding the conceptualisation of TVET and enhance relevant procedures for developing and comprehending curricula, training programme monitoring, instructional assessments and evaluating TVET programmes and provision.

TVET in Sierra Leone remains supply versus demand-driven, and therefore, its graduates do not have all the required skills relevant to the world of work. Employers expressed disappointment over the performance of TVET recruits, citing a lack of graduates in the occupations they need and graduates' overall lack of practical knowledge and experience.

Overall, understanding of quality assurance among NCVTA is focused on measuring inputs such as curricula, assessments and school infrastructure, instead of outputs like TVET graduate employment or whether employers are satisfied with the quality of TVET hires.

During a mapping exercise, almost all employers felt that they cannot influence the TVET system and that it is difficult to engage because they have a limited understanding of the system, and their inputs are sometimes not prioritised. Of the types of TVET employer engagement, the most frequently cited is on-the-job training for trainees. More robust policies and frameworks are not in place for this and all other types of Public-Private Partnerships (PPPs) in TVET. From the demand side, there still is no national labour market information system that would help detail occupation demand and ideally integrate with a MIS with TVET supply data. Finally, public awareness and understanding of TVET are limited, and there is no coordinated approach among providers to improve this.

Private Sector in PPP in TVET in Sierra Leone

The private sector can play a significant role in consultations on skills needs and the planning of vocational education skills in the workforce. The current skills demanded by employers do not precisely match the skills required by the market. This has resulted in a skills gap which comes at a cost for the economy (Carter, 2008; ILO, 2010). PPP in TVET is of particular importance since it can be comprehensively designed to encompass all aspects and levels of education and training, including financing, infrastructure, quality, in-service training of TVET instructors, establishment and review of vocational trades and technical courses. Skills development programmes necessarily involve close collaboration with industry and enterprises under the overall coordinating role of government.

PPP Framework: Private Sector Participation in Skills Development in Sierra Leone

Although the development needs of least developing countries (LDC) or a sector in Africa are similar, these countries and/or sectors are at varying stages of the development continuum. There is a difference in the specific development of these countries; hence, the PPP framework cannot be copied from one country to another country. There is no single model PPP framework that can be used across countries across a specific region. The PPP framework has been demonstrated as complex and filled with pitfalls and challenges (Chan et al., 2005). Therefore, to facilitate the success of the growing PPP infrastructure or development programme, researchers should continue to observe the PPP approach, identify potential issues and provide solutions to resolve these development challenges.

A PPP framework for a specific country would typically evolve over time, often in response to specific challenges facing the country and the implementation of its PPP programme. In the early stages of developing the PPP framework, the emphasis is on consulting with appropriate stakeholders to collect the necessary data enabling the consultant to create and promote PPP opportunities.

The components of a comprehensive PPP framework typically include the following:

- **Policy:** articulation of the rationale behind the intent to use PPPs to deliver TVET services, the objectives, scope and implementing principles of the PPP programme.
- **Legal framework:** the laws and regulations that underpin the PPP programme, i.e. enabling the partnership to enter into PPP agreement and setting up the rules and bound-

aries for how the PPPs are implemented. This can include PPP-specific legislation, laws and regulations or sector-specific laws and regulations.

- **Processes and institutional responsibilities:** the steps by which PPP projects are identified, developed, appraised, implemented and managed and the roles of different entities in the process. A sound PPP process is efficient, transparent, and followed consistently to control the quality of PPP projects effectively.

Establishing a PPP Framework

TVET Sector Analysis

Sierra Leone is grappling with high youth unemployment, poverty, inequality and slow economic growth. The unemployment situation is so severe that young people who graduate from tertiary and TVET institutions find it difficult to be employed. This is because they do not have the right skills needed by employers.

Therefore, there is an intention on the part of the government to put strategic emphasis on exploring the potential role of public-private partnerships in promoting alignment in skills development. The government is exploring the development of an integrated partnership approach in skills training within the TVET sector that will bring together all stakeholders, including public, private and non-profit, National Council for Technical, Vocational and other Academic Awards (NCVTA), National Council for Technical and Vocational Education (NCTVE), other related TVET departments as well as employers and employer organisations. This proposed framework could help to expand opportunities for work-integrated learning programme placements and increase the avenues for youths to find employment after graduation (Table 3.2).

Operationalising the PPP Framework in Sierra Leone

The operationalisation of the PPP framework will take the following:

Table 3.2 Roles and responsibilities of actors in key policy promoting, driving and supporting

Department of TVET Education	Promote partnerships through policy and regulation, provide funding and monitor partnerships in TVET (thus far, the focus has been on the number of partnerships)
TVET colleges and institutions	Form effective partnerships with industries for workplace-based placement, curriculum design and graduate employment
NCVTA	Facilitate partnerships with industry
	Coordinate PPPs at the sectoral/ regional level
	Promote the involvement of industry in curriculum development
	Articulate skills need in sectoral systems
	Fund skills development
	Promote and facilitate policy-related social dialogue and government accountability
National Skills Fund	Promote and fund partnerships
Sector skills council, committees at colleges and institutions	Facilitate partnerships between colleges, occupational/trade organisations and organised business
Public financing support employers	Fund skills development
	Articulate skills need
	Partner with colleges to advise on the curriculum, provide opportunities for practical training/ upskilling/career development of lecturers and provide workplace-based exposure opportunities to students

Step 1: Formation of a PPP-TVET Partnership Committee.

This step comprised the formation of a committee made up of representatives of key stakeholders, including the TVET coalition and the private sector, and chaired by the PPP Unit. The Unit's role will consist of policy formulation and development relating to PPP and TVET, providing an update on the legal framework that supports the PPP TVET process and clarifying and/

or contributing to the national economic guidelines concerning TVET.

Other stakeholders' committee members will include representatives of the trainers/teachers and students enrolled in TVET institutions. Mechanisms will be instituted to ensure adequate feedback from young people and instructors on the effective application of curricula, testing, training and employment uptake by graduates to assess the programme's success. Effective partnership building and sustaining a productive education-industry partnership require commitment, time, effort and involvement by all the stakeholders.

Step 2: Assign Roles and Responsibilities of Key Stakeholders.

The following roles will be assigned to each of the key stakeholders:

The Role of the Government

The government willingness to relinquish its monopoly on provision of training and enter into PPPs will be a crucial indication of its seriousness to move forward with the PPP process. The strength of the Government of Sierra Leone in demonstrating its commitment to the development of the PPP process would be reflected in its enthusiasm to develop a monitoring system and provide accreditation and certification and incentives to the other stakeholders to participate in the process through subsidies, tax waivers, inclusion in decision-making and support for critical values that lie outside the immediate tangible gain such as equity. As stated in the advisory note on PPPs in TVET, the following should also be considered by the government.

The Role of TVET Institutions

The role of TVET institutions in ensuring a successful implementation of the PPP framework is crucial. Their roles will include identifying the right private industry that would support the PPP framework and aid in the specified skills area. Their willingness to accept the standards of the industry and the examinations system is paramount to a successful PPP-TVET partnership. The TVET institution should also be willing to provide teachers and instructors and be responsive to in-service training that the industry will be providing.

Step 3: Identify Challenges in the Relevance and Quality of the TVET Courses, Skills Mismatch, Etc. and Solutions as Illustrated in the Table Below.

Step 3 is crucial to the overall development of the PPP process. During this step it is important that mechanisms such as refined survey instruments and monitoring and evaluation systems are put in place to ensure data collection and feedback through the PPL process. Here the organisers will have a mechanism which gives them direct contact with key stakeholders in TVET to enable them to collect data on all areas of the TVEP process including the relevance and quality of the TVET programmes.

Step 4: Selection of PPP-TVET Partnership Model That Can Be Considered in Sierra Leone Context.

The government will assess and select a PPP model from the list of the different models of PPP-TVET partnership models that can be considered in Sierra Leone context. This can be dealt with on a case-by-case basis which will involve the government, the industry and the specified TVET institution.

Step 5: Formation of a Binding Document That Specifies the Roles, Responsibilities and Mode of Operations of All Parties Concerned.

The relationship between the private sector and Government of Sierra Leone can be formalised by an MOU or other such legal provision. The agreement could take the form of an agreement with a particular sector institution to carry out training in a specific sector, and/or the other might be an agreement between a learning group and the government providing learning facilitators. The roles of the key stakeholders (the government, the private sector/industry and the TVET institutions) and the method of operation will be detailed in such a document.

Step 6: Data Analysis for Performance Indicators.

This partnership needs to be evaluated to assess the various indicators for success. These indicators include responsiveness, relevance, cost-effectiveness, efficiency, accountability and sustainability to measure effectiveness of PPP process. This process will enable the government and key stakeholders to evaluate the relevance of the courses, cost-effectiveness of the trainings, training methods, equipment and tools. The table below illustrates details of the performance indicators (Table 3.3).

Conclusions and Recommendations

Conclusion of the PPP Framework

Sierra Leone is faced with acute skill shortage. This represents a major challenge for the country as it tries to improve its global competitiveness. Therefore, despite government efforts, the Technical Vocational Education and Training (TVET) system remains complex, with multiple stakeholders operating at different levels and

Table 3.3 Details of the performance indicators

Performance indicators	Items
Relevance	Courses offered at training institutes are relevant (correspond to the industry needs) The equipment used during training is up to date Industries should be consulted periodically to provide specific advice on occupations and skills in demand and to identify key skills, project priorities for their sector The employability of TVET graduates because they have obtained a certain level of skills and technical competencies
Responsiveness	PPP for training institutes in TVET can produce more skilled students PPP for training institutes can produce graduates who meet the demands of the industry PPP remains a veritable tool to achieve national economic development PPP programme alerts of skilled training enhancement
Cost-effectiveness	PPP can ensure proper financial flows PPP help industry to overcome their financial constraints PPP will reduce administration costs that are borne by institutes/industry/government A specific percentage of income tax generated annually by the government should be utilised for the provision of training facilities are universities/training institutions in TVET Cost of operation is expected to be much lower due to the greater managerial efficiency in PPP Without investing a considerable amount of money upfront, government would be able to cater to a much larger student population. PPP will reduce costs and increase revenues
Efficiency	PPP shares cost savings PPP lowers bureaucratic constraints Decentralising roles, power and authority of PPP in TVET is important for effective delivery of training PPP promotes faster and more efficient delivery of services/information
Accountability	PPP contributes to high achievement in TVET training Private sector should be the key player in PPP activities/programmes Government is highly responsible for supporting PPP in terms of funding and training facilities provision Training institutions should take a lead in future planning with regard to PPP
Sustainability	PPPs in training institutions produce competitive graduates PPP in TVET is designed to produce the skilled and entrepreneurial workforce that is needed to create wealth and would help reduce the menace of poverty and unemployment PPP in TVET prepares persons for employment in a recognised occupation

moving into different directions, which has led to many inefficiencies and duplication of efforts. This has also led to the duplication of efforts and wastage of the limited resources available to drive the countries skills training and development process (Carter, 2008).

In recognition of the above situation in Sierra Leone, a fundamental shift is needed from quantity to quality in terms of its skills development process and more engagement with private sector to achieve a demand-led TVET system that answers to the development needs of the workforce and the private sector. The government committed to work with private sector agencies to accelerate changes in policies, including lowering taxes and providing incentives to encourage businesses to participate in the TVET schemes. It is recognised by the government that for the private sector to become competitive, they will be required quality human capital to enhance efficiency and quality of output. The PPL framework provides an excellent opportunity to the private sector to shape and influence the curriculum, content and mode of instruction of TVET institutions in terms of the skills and competencies their staff will acquire during training. The private sector agency will in turn provide financial support to sponsor trained, technical advice, modern tools, apprenticeship and internship opportunities to youth will pursuing their studies. This shift in development of a PPP framework will enable the government of Sierra Leone to address the mismatch on the skills that are provided by training and education institutions and those that are required by the market (ILO, 2010; Carter, 2008; Simmons, 2021; Government of Sierra Leone, 2019a, b).

Therefore, there exists significant space within the system for new players to be engaged in successful and innovative pathways in which the private sector can be engaged in the development and implementation of the TVET sector in Sierra Leone.

Appendix 3: Summary of the YEEP Pilot Model Framework

Introduction

Entrepreneurship is considered by the Government of Sierra Leone as one of the solutions to address high youth unemployment in the country, and as a result, there are several initiatives focused on the promotion of entrepreneurship among young people (Government of Sierra Leone, 2019a, b; IMF, 2019). The Government of Sierra Leone along with several stakeholders have launched numerous initiatives to promote self-employment, especially targeting the younger population. However, there is still a consensus that despite numerous governments and private-led initiatives, the system to support starting entrepreneurs remains weak, with the lack of effective capacity building services for entrepreneurs identified as a huge limitation.

It is believed that the technical assistance will play a crucial role in the achievement of sustainable development, by identifying youth entrepreneurs and businesses with growth potential and providing them with the tools to expand and develop sustainably. This component is in line with the consortium's ambition of private sector development for job creation and economic development among Sierra Leoneans especially Sierra Leonean youth. With the strong experience regarding youth development, entrepreneurship, incubation, accelerator programmes and business development support services, the YEEP team possess the capacity and competencies to tackle the youth employment challenge through entrepreneurship and technical assistance (TA).

Sierra Leone remains among the world's poorest countries according to the Human Development Index (UNDP, 2020). It also ranks 154th on the list of exporting countries in the world.[2] The country still suffers from the severe impact of civil war, which lasted for 11 years

[2] https://atlas.media.mit.edu/en/profile/country/sle/

from 1991 to 2002.[3] Additionally, the outbreak of the Ebola pandemic in 2014 caused significant damages to the country already weak socioeconomic systems.[4] The emergence of the COVID-19 pandemic contributed to significantly worsen the economic prospect of Sierra Leone.

Youth unemployment is widespread and is estimated at approximately 70% of youth being underemployed or unemployed. The private sector of Sierra Leone is very weak, with a significant portion of enterprises being small. Bank lending to the private sector is only 30% of their portfolio. For sectors such as agriculture, it is only 7% of their portfolio. Interest rate ranges from 25% to 30%, making it exceedingly difficult for businesses to expand even in cases where these businesses are well established.

Sierra Leone's *Agenda for Prosperity* commits it on the path of becoming an inclusive green, middle-income country by 2035. Sierra Leone uniquely defines green growth as developing infrastructure, energy and cities sustainably; managing renewable and non-renewable natural resources efficiently; and building resilience for the benefit of citizens. The impacts of climate change, environmental degradation and strong dependence on natural resources have driven the country to ensure a sustainable development pathway. This strategy will enhance Sierra Leone's commitment to achieving the Sustainable Development Goals (SDGs), which it currently lags behinds. In ensuring green growth, the Government of Sierra Leone has identified certain key sectors and intervention strategies that will lead to the desired growth to make it a middle-income economy by 2030, which investment in human capital through entrepreneurship is viewed as one of the main vehicles for driving this new wave of economic expansion.

SWOT Analysis

The consultant utilised SWOT analysis to further explore the strength and weaknesses of the employment and entrepreneurial environment in Sierra Leone. The analytical tool also provides opportunity for the consultant to investigate the opportunities which exist and the threads to be impacted on employment creation and enhancing and/or retarding sustainable livelihoods in Sierra Leone.

Sectors of Interest

The key sectors of interest include agribusiness, renewable energy, tourism, ICT and transportation. These sectors were chosen within the context of the country's development macro-strategy, sectors' contribution to the economy and its ability to offer mass employment for people of Sierra Leone, and its synergies with other initiatives within the country.

Agribusiness and Fisheries

Sierra Leone's economy is largely dependent on agriculture. It contributed about 47–50% of the GDP. The crops that require investments are rice, cocoa, coffee, oil palm, cassava and vegetables.[5] A commodity like rice contributes to 75% of agricultural GDP. The supportive agro-ecological climate in Sierra Leone is a known fact. The total arable land is estimated at 4.3 million hectares,[6] with less than 20% having been put under cultivation.[7] Despite abundance of land which support rice cultivation in the year 2016, the country spent $76 million on rice importation. The importation of rice currently is estimate at $100 million.[8]

Opportunities within the agricultural sector run from production to processing. There are two major setbacks in the sector. These is the absence in the sector of measures, including a) technologies to enhance sustainable production and b) limited opportunities for ago-processing to add value to primary products.

In the level of fisheries subsector, the scoping and stakeholder interactions revealed a strategic

[3] https://en.wikipedia.org/wiki/Sierra_Leone_Civil_War

[4] https://www.who.int/csr/disease/ebola/one-year-report/sierra-leone/en/

[5] http://sliepa.org/investment/agriculture/

[6] file:///C:/Users/user/Downloads/Sierra%20Leone%20investment%20agriculture%20(1).pdf

[7] https://awoko.org/2016/10/18/sierra-leone-news-salone-spends-over-100m-to-import-rice/

[8] https://atlas.media.mit.edu/en/profile/country/sle/

interest by the Government of Sierra Leone to create sustainable jobs for youth in this sector. Unfortunately, this has failed due to a) poor business services in the fishery enterprises and b) the poor access to modern technologies.

Renewable Energy

Although Sierra Leone has abundant natural resources for significant energy production, access to energy is extremely low. Access to electricity is a challenge for most citizens in Sierra Leone. It is estimated that only 13% of the population have access to electricity. Other estimations from outlets such as the International Energy Agency estimate that less than 10% of Sierra Leoneans have access to electricity.[9] This electricity connection is consumed by the four main cities.

It is important to note that of Sierra Leone's 90 MW of installed capacity, 86% is consumed mainly by Freetown (the capital city). In the rural areas, less than 2% of the population have access to electricity.[10] They often gain energy through biomass such as charcoal and firewood and diesel-powered generators. It is estimated that there are 33,000 generators currently in use that provide a capacity of approximately 180 MW.[11] Sierra Leone is currently experiencing a net deficit of power with average peak demand requirements of 300–500 MW. The country is in desperate need of new power and renewable sources.

The Government of Sierra Leones is seeking to develop cheap, affordable energy for all of its citizens by leveraging its renewable energy (RE) resources. Solar energy systems have already proven itself as cheap alternative energy source for Sierra Leoneans. Additionally, hydropowered energy systems have been identified as a critical potential source of energy for multiple applications.

Tourism

Sierra Leone as a least developed state in West Africa offers bountiful tourism opportunities. These include bird watching, sport fishing and climbing in nature preserves such as Outamba-Kilimi National Park, Lake Sonfon, Gola forest and Tiwai island sanctuary. The tourism sector also includes 350 kilometres of prime beaches, untapped unique landscape, scenery and cultural heritage. In 2012, revenue generated from the tourism industry stood at $41.6 million, and in 2013, revenue generated $58.6, which shows a progressive growth in the industry.

The potential in the sector offers significant opportunities for businesses providing all kinds of services to tourists. This could include ticketing, transportation, hotel/lodging, tour guides, restaurants and many more. In 2016, the total contribution of travel and tourism to employment, including jobs indirectly supported by the industry, was 3.2% of total employment (41,000 jobs). This is expected to rise by 2.7% pa to 58,000 jobs in 2027 (3.4% of total).[12]

Information Communication and Technology (ICT)

Information communication and technology (ICT) opportunities are abundant since the current uptake of these ICT technologies is within an exceedingly small fraction of the population. For example, less than 50% of the population have access to mobile phones.[13] Mobile money services are still of exceptionally low patronage. However, it is estimated that over 200,000 people in West Africa, predominantly in the retailing and distribution of services and handsets, gained employment from mobile operators and the wider mobile ecosystem in 2017.[14] The country is in the process of attracting high-level investments into the ICT sector that will ultimately stimulate

[9] https://awoko.org/2018/04/24/sierra-leone-news-less-than-10-of-sierra-leone-has-access-to-electricity/

[10] https://energypedia.info/wiki/Sierra_Leone_Energy_Situation

[11] https://www.climateinvestmentfunds.org/sites/cif_enc/files/meeting-documents/sierra_leone_eoi_0.pdf

[12] https://www.wttc.org/-/media/files/reports/economic-impact-research/countries-2017/sierraleone2017.pdf

[13] https://www.ifc.org/wps/wcm/connect/region__ext_content/regions/subsaharan+africa/advisory+services/accessfinance/partnership_financialinclusion/scopingreport_sierraleone

[14] https://awoko.org/2018/06/13/sierra-leone-news-55-of-sierra-leoneans-do-not-use-mobile-phones/

downstream activities among SMEs. These investments include Africa Coast to Europe (ACE) submarine fibre cable – a $31 million loan facility from the World Bank to improve speed of Internet and enhance associated services related to voice, data and video.

Target Groups

The Graduate Entrepreneurship Programmes (GEP) primarily focus on early-stage businesses established and run by graduate youth. The GEP will be engaged with young entrepreneurs in line with the definition of youth in Sierra Leone. These businesses should contribute to the development of the green economy of Sierra Leone, and thus they must address at least one of the four sectors of interest. The businesses in question should require financing (credit) between $3000 and $10,000.

The businesses should have potential to create employment opportunities for both skilled and unskilled youth (including women). Businesses whose operations have linkages with economy of rural communities such as agro processing and ecotourism will be given priority. Some common features among the target businesses are lack of collateral to enhance access to loans and the limited deal flow. The market in Sierra Leone is relatively smaller, and most entrepreneurs lack clear insight on market opportunities.

Aim and Objective of the Project

The YEEP is aimed at contributing to the Government of Sierra Leone's efforts through the National Youth Commission (Implementing agency) to create decent jobs and sustainable incomes for the youth. The overall aim of the project is to reinforce the employability of Sierra Leone Youth as well as enhancing their entrepreneurship skills through the development of a framework which integrates education and vocational institutions with job market demand and supply.

It is envisaged that this approach will result in strengthening the occupational levels in the youth population segments of the country which will eventually trigger sustained economic development in the growth sectors.

The objectives of the project are to:

- Improve the skills for employment of Leonean Youth.
- Strengthen the capacities of key stakeholders and institutions involved in youth development, entrepreneurship and employment.
- Scale up successful pilot initiatives.

As this is a pilot project and it is expected that the approaches and methodologies developed will be used to scale up, institutionalise and sustain, this conceptual framework for the pilot model has been developed. To fully contribute to the general objective of strengthening entrepreneurship and creating employment for young people through the development of models, capacities and systems linking the education/vocational institutions with the market, the assignment will in particular:

- Adopt an **overall piloting approach** characterised by the **modulisation** and the **measure** of its activities and results to secure its **replicability** and its **scalability**.
- Provide **high-quality tailored services** in the implementation of the activities.
- Focus on the **sustainability** of the activities and the outcomes and the **capacity building** of implementing institutions (government, education institutions, private sector, PIU …) and participants (TVET instructors, counsellors, BSC staff …).
- Promote **innovation** in the design and experimentation of the activities of the assignment, in particular the development of the Public-Private Partnership (PPP) framework and of the curricula as well as the establishment of the platform and of the networking.

Management and Coordination of the Project

The YEEP is implemented by the Government of Sierra Leone through the National Youth Commission (NAYCOM) and funded by the African Development Bank (AfDB). The YEEP is managed by a consortium headed by the Lattanzio Advisory SpA (Milan, Italy) and include Fondazione Giacomo Brodolini (Rome, Italy) and Fondazione AVSI (Milan, Italy/ Freetown, Sierra Leone).

Theoretical Framework of the YEEP Guiding Principles for Implementation

The methodology used in the pilot implementation is aligned to the guidance provided in consultation with stakeholders, lessons learned and recommendations from previous research done on youth the impact of climate change on youth in small island communities which led to the development of the Youth Climate Change Adaptation Framework (Simmons, 2021). It also considers the timeframe, scope and the intensity of project design. In undertaking the pilot implementation, the team technical experts comprised of the team leader, the key experts and non-key experts employed a step-like approach to activate the framework.

The implementation team worked towards improvement of systems and skills for systematic reflection on and learning from practice. The team brought new methodologies and tools to nurture the development of the institutional capacity of the National Youth Commission and the partner agencies participating in the project. The capacity development sessions for participating organisations focused on enhancing social skills such as effective communication, network dynamics, participatory decision making and gender awareness in a highly interactive manner, which contributed to the spread of knowledge.

Monitoring and Evaluation

All training courses will be monitored and evaluated to capture the views and perception of the participants, instructors and administrative officials of partner agencies on how effective and efficient the training courses are delivered. This is done via a confidential evaluation form as well as verbally monthly, mid-term and at the end of the training project. Positive aspects and suggestions for improvement are captured from participants, instructor, management of partner institutions and officials of NAYCOM. Evaluation results of all project activities will be archived and aggregated by a consultant and shared with the M&E specialist of NAYCOM. Results will be used to make improvement and continuous adaptation of the project.

The team would use the data from the M&E activities and physical visits to partner institutions to prepare monthly performance management and operational reports for submission to Lattanzio KIBs for submission to NAYCOM. It will also prepare mid-term and end of project reports for submission to NAYCOM. These reports will also bear in mind the various stages and deliverables of the project.

Monitoring Mechanisms for Each Component

Component 0: Project Management

This component of the project will be managed by the Project Lead Consultant of the YEEP on the following primary indicators:

- Approved monitoring framework.
- Submission of monthly reports.
- Number of capacity building sessions conducted for PIU.
- Approved strategy for upscaling and sustaining project.

Component 1: Skills Development for Employment

The Skills Development for Employment component will be managed by the key expert with responsibility for this aspect of the project. The following monitoring and evaluation strategy will be used for the project:

Measure the progress and quality of the outputs and activities and risks related to its implementation, such as the monitoring of key indicators and the mapping of risks.

It will be an ongoing process, and, if required, corrective measures will be taken at the earliest possible stage to avoid the accumulation of quality deficiencies that may be hard to remedy at a later stage.

The component will be monitored based on the following primary indicators:

- Successful completion of the PPP framework.
- Updated course content and training methodologies for the three priority skills and trades in selected TVET institutions.
- Complete the preparation of the entrepreneurship curriculum for the selected TVETs.
- Completion of the Training of Trainers for 20 TVET instructors for the priority skills and trades and entrepreneurship training.
- Completion of the training of 100 youths in supervisory skills.
- Completion of the training of 150 youth in the selected priority skills.

Furthermore, the following additional indicators were used to assess the progresses made in various output areas:

- Improved employment rate of youth participating in TVET training programme.
- Number of female beneficiaries participating in the project.
- Percentage of youths linked with private sector employers for employment.
- Number of capacity building sessions conducted.
- Increase in the number of inquiries and enrolment in TVET institutions.
- Number of communication activities conducted (e.g. PR events, press releases, conferences held, email alerts, etc.)
- Number of donor acknowledgements done via different media (e.g. banners, websites, press releases, etc.)

Operationalisation of the Skills Development for Employment Theory of Change Continuum Model

The consortium embarked on a comprehensive study on the status of youth employment in Sierra Leone using appropriate research methodology and tools. Skills development is critical to youth employments and job creation. The study discovered that youth in Sierra Leone have inappropriate skills to enhance their employability. This is due to the training provided to youth is not relevant to the needs of the job market in Sierra Leone (Carter, 2008, ILO, 2010). This outcome is also due to the non-availability of appropriate tools and poor quality of training provided by TVET institutions. It also discovered that TVET institutions have the infrastructure to carry out the technical trainings, but do not have modern equipment and nor tools to train students in the respective skills offered. The curricula implemented in these institutions were irrelevant to the needs of the market and do not consider the impact of climate change and other development calamities on Sierra Leone now and in the future.

Partnership was forged with the TVET secretariat and TVET institutions to develop the PPP framework to encourage private sector investment in TVET.

Curriculum and Modular Development

The skills development curricula was prepared by the consortium with inputs from key experts and institutions and was sent to NCTVA for review and approval. The contents include the three skills areas executed by component 1 of the YEEP including Sustainable Hospitalities (Hotel Management and Tourism), Sustainable Construction (Carpentry, Masonry and Aluminium Fabrication (a new course introduced at GTI) and Sustainable Agriculture (Aquaculture/Fish Farming). These curricula are designed to provide the appropriate guide and pedagogy for training students in partner institutions in order to enhance their skills development to help reduce the skills gaps in the labour market and promote youth employability. The curricula contain training guidelines, methodology and supporting materials for trainers/tutors to use in the execution of the specific skills training. The content for the training courses is sector specific, and the goal is to enable lecturers, instructors and supervisors to facilitate effective learning with positive results.

Identify Institutions and Signing of MOU

Partner institutions were identified followed by institutional assessment. Three (3) higher institutions (Milton Margai College of Education and Technology (MMCET) – Brookfields Campus was selected to deliver Hotel Management and Tourism course to TVET students, the Government Technical Institute - Ferry Junction for Construction and Aluminium Fabrication and Njala University for Aquaculture and Fish Farming) were identified by the consortium for partnership. Memoranda of Understanding (MoU) was developed and signed by representatives of partner institutions for the implementation of activities of component 1 of the YEEP project in the respective institutions based on a) availability of expertise to deliver the training and b) adequacy of training facilities/space and methodologies. The curricula then informed the development of skills-specific training manuals that institutions are using to train their tutors to deliver quality hands-on training to TVET students. Training modules for the 3 skills areas were developed, and 60 TVET instructors trained deliver competency-based trainings in the three partner institutions.

Training of Supervisors/Mentors and Students (Methodology, Pedagogy; Theoretical and Practical)

Trainings were conducted for TVET trainers/tutors in each of the partner institution using appropriate methodology and pedagogy. The curriculum that informed those trainings and practical sessions was central during the trainings of the supervisors and mentors. Facilitators or tutors were drawn from the private sector and the institutions considering their expertise in the various sectors (Aquaculture, Construction and Hotel Management and Tourism). The three separate trainings lasted for a period of 4 days each followed by certification of participants by NAYCOM.

Students were recruited in each institution from their specific departments followed by training in subsections involved in component 1 of the YEEP project. Trained lecturers and supervisors provided trainings and mentorships for TVET students. Their session was a blend of theory and practical, allowing them to gain fullest understanding and techniques of the various skills involved in each component. Trainings lasted for a period of 3–9 months, and this was followed by internship and job placement in the private sector. One hundred four youths (M: 61; F: 43) were trained on management and supervisory skills from the three institutions (Njala, MMCET & GTI). The trainings enhanced their mentorship and supervisory skills; hence the PPP framework is pivotal in this situation as it enables the private sector to integrate professionally trained students in various skills areas.

Employment and Employability

Once students have gotten the requisite trainings and gone on internships and job placements, they are fully equipped and ready to take-up decent jobs in their various fields. MOU was signed between MIRO Timber Company and GTI for transfer of knowledge, provision of technical and material support to the TVET Institution and job placement. A series of meetings were held with hotels and restaurants and travel and tourist agencies for internship opportunities for students from MMCET. Students from the Aquaculture and Fisheries Management Department have been placed on internship at Agro Fish-Farm in Bo, Njala University Fish Farm at Mokonde and Mano Daise in Moyamba District Company in Bo. Other private sector players have been contacted to provide partnership and internship/placement opportunities to partner institutions. The COVID-19 pandemic impacted severely on the number of students, specifically from the Sustainable Hospitalities and Sustainable Construction courses who were able to take up internship/placement opportunities.

Component 2: Career Guidance and Job Readiness

The deliverables will be monitored through reports of activities undertaken during the YEEP implementation process. The trainings will be monitored using training evaluation forms including:

- Satisfaction questionnaires for each training course, including trainers' performance.
- Questionnaires for the evaluation of the environment in the classroom.
- Final test for the assessment of the knowledge and skills acquired (if requested).
- Elements of the training evaluation will include:
- Training objectives.
- Methods used by trainers.
- Relevance of the presentation.
- Course content.
- Venue.
- The main output indicators for which component 2 will be monitored are under three key deliverables:
- Career guidance and job readiness programme.
- 30 counsellors trained in designed Career Guidance and Job Readiness programme
- 2400 youth targeted in selected TVET institutions and universities and provided with training in career guidance and counselling as a pilot
- Further indicators to highlight progress made for each output indicator include:
- Number of female beneficiaries.

Operationalisation of the Career Guidance and Job Readiness Theory of Change Continuum Model

Research on Employment and PPP Framework

Research on Career Guidance and Job Readiness was also conducted using similar research methodology as in the case of Skills Development for Employment. The study findings were thousands of graduates mistakenly choose courses in the university that do not match the expectations of the job markets. The result of this situation often leaves them being unemployed after graduation from the university. Hence, career guidance and counselling are very important to support students choose courses of their choice that will help them attain secured and decent employments after graduation. Career Guidance and Job Readiness (component 2 of the YEEP project) is therefore appropriate to help address these growing issues affecting our young graduates in almost all the universities in Sierra Leone.

The Career Guidance and Job Readiness programme has been approved and now institutionalised in all partner institutions across the country. The programme is now accessible by students in the various institutions: SOS, MMCET – Goderich Campus, EBK, Eastern Polytechnic, GTI and Njala University (Bo and Mokonde Campuses). Sixty (60) counsellors/university lecturers were trained on Career Guidance and Job Readiness modules for a possible cascading to the targeted 2400 students.

Curriculum and Modular Development

The Career Guidance and Job Readiness curriculum and training manuals were developed by Career Guidance and Job Readiness experts. The curriculum holds the training methodology and topics to be covered during the actual training of students. On average 80% training completed on all campus. 100% training modules have been uploaded on the platforms for counsellors to access and use as a guide when teaching.

Identify Institutions and Signing of MOU

Four partner institutions were identified based on the establishment of CAPS centres in those institutions and the availability of structures and facilities. Memorandum of Understanding (MOU) was signed to establish relationship and pathway for the implementation of component 2 activities of the YEEP project. Due to high interest demonstrated by youths across the country in the career programme, other institutions, SOS, MMCET, EBK and Njala University, were enlisted to deliver Career Guidance and Job Readiness courses on the campuses. The inclusion of the four institutions and the e-learning technology helped the career programme to record a total of 6130 students (255% more that proposed target) undergoing training on the different campuses across the country.

Training of Supervisors and Students (Methodology, Pedagogy; Theoretical and Practical)

Training of Trainers sessions were conducted for lecturers, counsellors and staff attached to the CAP centres of the partner institutions. The trainings were guided by the curriculum developed for Career Guidance and Job Readiness. Two 4-day trainings of trainers (ToT) for counsellors per institution was conducted in Bo and Freetown, respectively. A total of 60 counsellors were trained on Career Guidance and Job Readiness modules.

Also, students were recruited in each partner institution and trained by counsellors, mentors and lecturers. Trainings were student-centred, and practical counselling sessions on career development were conducted too. Students' training lasted for a period of 3–9 months. E-learning was developed to provide the opportunity for the continuation of the training and interaction between students, counsellors and lecturers during the COVID pandemic when social distancing and other preventive measures were enforced.

Employment

The graduating students are believed to have made appropriate choices on their career paths that will earn them sustainable employment in both the public and private sectors.

Component 3: Graduate Entrepreneurship Programme

The project result framework has identified the following project outcome indicators:

- Approved training manuals and operational instructions.
- Approved entrepreneurship training programme.
- Number of youths trained in entrepreneurship.
- Number of counsellors and mentors confirmed on business counselling and mentoring network.
- Number of youths active on entrepreneurship platform.
- Specific indicators to monitor include but not limited to:
- Increase in number of businesses started by youth participating in programme.
- Number of female beneficiaries.
- The main output indicators for which component 3 will be monitored are under three key deliverables:
- 16 staff members of BSCs trained
- 105 young people trained in entrepreneurship development
- 60 young people incubated and accelerated.

Operationalisation of the Graduate Entrepreneurship Theory of Change Continuum Model

The development of the pilot model of the conceptual framework for the Graduate Entrepreneurship Programme is informed by the outcomes of the SWOT analysis to assess the entrepreneurial environment in Sierra Leone. The pilot model framework is aligned with the critical issues surrounding the entrepreneurial growth of youth in Sierra Leone. This is a transformational process focusing on building sustainable youth-led business and employment in Sierra Leone. The process starts with the recruitment stage that provides the platform for graduate entrepreneurs to access the opportunities to be trained and mentored in business development and entrepreneurship. The weakness in the SWOT analysis highlights the growing gap in the entrepreneurial ecosystem in Sierra Leone. However, such weaknesses can be transformed through the GEP intervention by enabling the graduate youths to develop and establish sustainable businesses with the ease of accessing finance to support the development and expansion of youth businesses.

Achieving the GEP goal requires graduate entrepreneurs to go through a process that is highlighted in the pilot model conceptual framework. From the ideation stage to the acceleration, the graduate entrepreneurs are encouraged and guided to develop bankable business ideas. The ideation stage is focused on equipping the youth entrepreneur with business strategy using the business model canvas as a principal tool. Coaching forms part of this process as it enables entrepreneurs and mentors to discuss critical issues in their product development.

At the incubation stage, three thematic areas are emphasised for keen attention: market research, operations and human resource and financial modelling. The graduate entrepreneurs are subjected to get out of the training facility to reach their potential customers (conduct surveys) for feedback on their product to inform the development of their marketing strategy and competitive analysis. Operations and team enable the graduate to organise the business in a manner that is less harmful and recruits the appropriate personnel to run the business. Graduate entrepreneurs are encouraged to develop a suitable financial model that enables them to understand the viability of their businesses.

The acceleration stage prepares the graduate entrepreneurs on managerial competence, operational ability and growth potential of entrepreneurial ventures. The stage enables graduate entrepreneurs to build business partnership and attract investors and bankers through pitch development and presentation. Coaching and mentorship are critical in this stage. Thus, graduate entrepreneurs are assigned to key mentors with specialty in their respective businesses.

Summary of the YEEP Pilot Model Framework

The YEEP project was instituted to solve the problem of youth unemployment through enhancing the entrepreneurship culture among youth people in Sierra Leone. The YEEP comprised three components including component 1 which concerns with provision of skills training to enhance the employability of youth, component 2 which comprises of the provision of Career Guidance and Job Readiness training to youth in tertiary and TVET institutions to enhance their career choice by providing soft skills to enhance their employment prospects and component 3 which intends to inculcating a culture of entrepreneurship among graduate youths through training in entrepreneurship to equip them with skills to establish their own businesses for creating jobs for themselves and others.

The YEEP adopted Youth Climate Change Adaptation Framework (Simmons, 2021) which concerns with development of the skills and knowledge of young people to adapt to building their resilience to the compact of climate change and creating sustainable livelihoods via entrepreneurship. The YEEP encompasses adult learning methodological approaches that are linked to the theory of change conceptual framework to address youth unemployment and poverty in Sierra Leone. This approach aimed at creating sustainable youth-led enterprises across the country. Institutional assessments and Training of Trainers' initiatives were organised to equip partner institutions to implement the YEEP within their respective districts/regions.

Overall, the goal of YEEP is to provide support to youth and the training institutions to provide skills and knowledge to youth to enable them to establish sustainable enterprises, create jobs and become self-employed. The YEEP enhance youth employability from a practical perspective by addressing the development challenges discussed in the background of the paper. Through the YEEP, it is believed that youth will be equipped with the skills, knowledge and competences to establish their own enterprises and create job opportunities for themselves and others and contribute significantly to building a sustainable economy in Sierra Leone.

References

Angelsen, A., Larsen, H. O., Lund, J. F., Smith-Hall, C., & Wunder, S. (2011). *Measuring livelihoods and environmental dependence; methods for research and fieldwork*. Earthscan.

Bradshaw, Corey, J. A., Ehrlich, P. R., Beattie, A., Ceballos, G., Crist, E., Diamond, J., Dirzo, R., Ehlich, A. H., Rodolfo, M. E., Pyke, G., Raven, P. H., Ripple, W. J., Saltre, F., Turnbull, C., Wackernagel, M., & Blumstein, D. T. (2021, January 13). Underestimating the challenges of avoiding a ghastly future. *Frontiers in Conservation Science, 1*, 1–10.

Carter, R. (2008). *Caribbean youth: An integrated literature review*. Bridgetown, Barbados.

Chan, A. P. C., Lam, P. T. I., Chan, D. W. M., Cheung, E., & Ke, Y. (2010). Critical success factors for PPPs in infrastructure developments: Chinese perspective. *Journal of Construction Engineering and Management, 136*(5), 484–494.

Chan, W. T., Chen, C., Messner, J. I., & Chua, D. K. (2005). Interface management for China's build-operate-transfer projects. *Journal of Construction Engineering and Management, 131*(6), 645–655.

Fengyu, B., Chan, A. P. C., Chen, C., & Darko, A. (2020). Review of public-private partnership literature from a project lifecycle perspective; journal of infrastructure systems. *Journal of Infrastructure Systems, 24*(3), 1–13.

Government of Sierra Leone. (2004). *Report on the review of education act*. Government of Sierra Leone.

Government of Sierra Leone. (2015). *Population and housing census thematic report on economic characteristics*. Government of Sierra Leone.

Government of Sierra Leone. (2019a). *The Government of Sierra Leone's new medium-term National Development Plan (MTNDP) 2019–2023*. Government of Sierra Leone.

Government of Sierra Leone. (2019b). *Youth skills development for entrepreneurship and employment, 2019, National Youth Commission (NAYCOM)*. Government of Sierra Leone.

Harley, M., & van Minnen, J. (2010). *Adaptation indicators for biodiversity* (European topic centre on air and climate change technical paper 2010/15). European Environment Agency. https://nctva.org/about-us/

ILO. (2010, November). *A skilled workforce for sustainable growth: A G20 training strategy*. International Labour Office.

ILO. (2013). *Sustainable development, decent work and green jobs*. International Labour Conference, 102nd Session, International Labour Office, Geneva.

IMF (2019) *Sierra leone: Economic development documents- national development plan 2019–2023: IMF staff country reports*, International Monetary Fund Africa Department. 9781498324960/1934-7685.

IPCC. (2014a). Climate change (2014) synthesis report. In *Contribution of working groups I, II and III to the fifth assessment report of the intergovernmental panel on climate change* (p. 151). IPCC.

IPCC. (2014b). Summary for policymakers. In C. B. Field, V. R. Barros, D. Dokken, K. J. Mach, M. D. Mastrandrea, T. E. Bilir, M. Chatterjee, K. L. Ebi, & Y.O. Estrada (Eds.), *Climate change impacts, adaptation and vulnerability* (Contributions of the Working Group II to the fifth assessment report). IPCC, Geneva.

Jaja, J., Dawson, J., & Gaudet, J. (2016). *Using social network analysis to examine the role that institutional integration plays in community based adaptive capacity to climate change in Caribbean small island communities*. Department of Geography, Environment and Geomatic, University of Ottawa, Canada: Routledge, Local Environment.

Kruss, G., Petersen, H., Fongwa, S., Tele, A., & Rust, J. (2017). *Synthetic analysis on the skills development and economic responsiveness role of education and training Institutions in South Africa: Towards an integrated public-private partnership strategy for skills development in the TVET college system: LMIP Project: A study of labour market interactive capabilities*. Structures and Mechanisms in Diverse Post-School Education and Training Institutional Settings, South Africa.

Margolis, D., Rosas, N., Turay, A., & Turay, S. (2016). *Findings from the 2014 labor force survey in Sierra Leone* (World Bank studies). World Bank.

Matland, R. (1995). Synthesising the implementation literature: The ambiguity-conflict model of policy implementation. *Journal of Public Administration Research and Theory, 5*(2), 145–174.

Nurse, L. A., McLean, R. F., Agard, J., Briguglio, L. P., Duvat-Magnan, V., Pelesikoti, N., Tompkins, E., & Webb, A. (2014). Small islands. In V. R. Barros, C. B. Field, D. J. Dokken, M. D. Mastrandrea, K. J. Mach, T. E. Bilir, M. Chatterjee, K. L. Ebi, Y. O. Estrada, R. C. Genova, B. Girma, E. S. Kissel, A. N. Levy, S. MacCracken, P. R. Mastrandrea, & L. L. White (Eds.), *Climate change 2014: Impacts, adaptation, and vulnerability. Part B: Regional aspects. Contribution of working group II to the fifth assessment report of the intergovernmental panel on climate change* (pp. 1613–1654). Cambridge University Press.

OECD. (2013). *The nature of policy change and implementation: A review of different theoretical approaches*. Organisation for Economic Cooperation and Development.

Pereznieto, P., Gbedemah, C., Rosenem, G., Haper, C., & Jones, N. (2011). *Youth vulnerability and adaptation: Exploring the impact of macro-level shocks on youth 3F crisis and climate changes*. Overseas Development Institution.

Poschen, P. (2015). *Decent work, green jobs and the sustainable economy: Solution for climate change development*. International Labour Organisation.

Ripple, W. J., Wolf, C., Newsome, T. M., Barnard, P., & Moomaw, W. R. (2019). World Scientists' warning of a climate emergency. *Bioscience, 70*(1), 8–12.

Simmons, A. (2021). *Impact of climate change on youth in small island communities Vol 1* (Sustainable development goals series, climate change). Springer Nature, Switzerland.

Stochowiak, S., Robles, L., Habtermanion, E., & Maltry, M. (2016). *Beyond win: Pathway for policy implementation*. ORS Impact/The Atlas Learning Project.

Thuraya, K. (2015). *Improving the effectiveness of E-learning improvement in the school of engineering at Tripoli University*; Doctoral thesis, University of Huddersfield.

UNDP. (2015). *Human development report*. Work for Human Development, United National Development Programme (UNDP), New York.

UNDP. (2020). *Human development report: The next frontier, human development and the Anthropocene*. UNDP. Available at: http://report.hdr.undp.org/part-1.html.

UNEP. (2021). *GEO-6 for Youth*. UN Environment Programme. Kenya, Nairobi.

Western Phoebe. (2021). *Scientist warns of ghostly future of mass extension and climate disruptions*; The Guardian Newspaper: The Guardian, 13th January 2021.

World Bank. (2017). *Sierra Leone systematic country diagnostic*. World Bank.

Xiong, W., Chen, B., Wang, H., & Zhu, D. (2018). Governing public-private partnerships: A systematic review of case study literature. *Research and Evaluation: Australian Public Administration, 78*(1), 95–112.

4 The Assessment Tools for Boosting Organisational Capacity and Efficiency to Implement Climate Change Adaptation Initiatives

Abstract

Section 1 of the chapter discusses the use of assessment tools such as the Cultural Web, to assess the culture in NAYCOM. The cultural web includes routines and rituals, power structures, organisational structure, symbols, stories, and paradigms. The consortium utilised the Information Technologies and Information System (IT/IS) to review how IT is used to support management's strategic decisions to enhance the project's overall operationalization. Section 2 discusses the challenges, methodologies, management, operations, and the achievement of the project implementation process deliverables. It shows that organisations should focus on applying technology, culture, and BPR strategies to enhance decision-making support, operational mechanism, and ultimately, competitive advantage.

Keywords

Cultural web · Information technology · Climate change · Sustainable development · Youth development · Entrepreneurship · Employment · Change management · Climate change adaptation · Africa · Sierra Leone · Biodiversity · Technical vocational education training · Conceptual framework

Introduction

Favouritism; nepotism, including tribal affiliations; and other forms of patronage are factors of undue influence in selection, training and promotion in Sierra Leone, even in the civil service. Therefore, to successfully implement a change process, change managers must implement appropriate strategies to address the service's culture effectively. Deal and Kennedy (1982) in Burnes (2000) argue that a strong culture is a powerful lever to guide change and define business culture as encompassing how people in a company are likely to act in each situation both inside and outside the organisation. Cultural theorists such as Hofstede believe that cultural influences play a critical role in the way employees behave in an organisation and the introduction of technologies only produce insignificant comparison (Beardwell & Holden, 1997). These assessment tools were used to gain insights into the working of the organisation, identify gaps and devise strategies to enhance the capacity of the organisation to deliver the YEEP and mitigate organisational pitfalls.

NAYCOM Cultural Web

The cultural web can be used to analyse the culture in the National Youth Commission of the Sierra Leonean government in that it examines

A. Simmons, *Climate Change Adaptation Framework and Youth Entrepreneurship in West Africa*, Sustainable Development Goals Series, https://doi.org/10.1007/978-3-030-85754-7_4

how organisations behave and change (Johnson & Scholes, 2002; Johnson et al., 2017). The National Youth Commission is a governmental public service agency in Sierra Leone that specifically addresses Sierra Leonean youth issues. The NYC can be defined as having a role culture that is bureaucratic, conservative, mechanistic, rigid and hierarchical, where adherence to authority is emphasised (Handy, 1986). Staff stick rigidly to the narrow tasks outlined in their job description rather than taking the initiative to address issues that appear challenging or outside of their work scope.

The NAYCOM culture presents the following features as indicated by Johnson and Scholes (2002) cultural web:

Stories

Stories emphasise the negative experiences of persons who believe that they have been failed by the organisation or are villains who have benefited from the services through improper and dishonest means; the focus is on individualism and not teamwork. Stories also demonstrate how political affiliation and personal networking will guarantee promotion and job security. The politicians are viewed as the villains by these civil servants and the public alike. Yet, civil servants still strive to be associated with them to secure their tenure in the workplace and acquire a promotion or other special favours, often to the detriment of those who wish to work honestly and conscientiously. The stories promote messages that discourage individuals from doing an honest job for honest pay.

Routine and Rituals

Employees are concerned with performing their day-to-day chores and feel disparaged to think strategically about solving long-term challenges. A strong blame culture is present in the civil service in Sierra Leone. The ritual concerns middle and senior managers assigning blame to lower-ranked employees for not achieving outcomes. Recruitment is mainly through friendship, personal and tribal networks and political affiliation, limiting the potential access to diverse workforces. Hoarding or gatekeeping information by middle managers and some junior staff has become a new ritual.

Symbols

Most junior employees aspire to be promoted and acquire the large private office and desk enjoyed by middle and senior managers. Seniority is an important status symbol in the public service in Sierra Leone, having implications on office size, parking space, opportunities to have a duty-free vehicle, their telephone bill paid by the organisation and other tangible benefits. Employees view the civil service as an institution where jobs are provided for life.

Power Structure

Power is mainly concentrated in the hands of officials, most of whom are affiliated with the ruling political party. Officials who possess the expertise and have specific tribal and family connections share a more significant power margin.

Organisational Structure

The public service is viewed as a rigid, conservative and hierarchical organisation. It possesses a clear pecking order among ministries and departments. Ministries and departments are given prominence depending on their revenue generation and collection roles within the civil service; the Ministry of Youth and NAYCOM are low on the order. There is evidence of tribalism among officials in line ministries and departments for recognition and budgetary support. Information and power flow are top-down; there is little evidence of bottom-up information flow (Matland, 1995; OECD, 2013).

Control

The auditor and financial reports are the central systems of control in the civil service in Sierra Leone. Some ministries use confidential report to penalise employees who deviate from the organisation's norms. The exertion of power by senior managers over staff is a crucial feature of the organisation's culture. Recruitment, selection, orientation, placement and appointment procedures are structured to maintain a cycle of dependency rather than self-direction, leading to laziness, a lack of drive and low-performance levels. This poor attitude is perpetuated throughout the NAYCOM and the entire civil service.

Paradigm

Civil servants view NAYCOM and the overall civil service as an appropriate government body delivering free services to persons who require such service or as a chaotic institution that provides some services below acceptable standards by public members. There is a poor image of the civil service within the country; public servants' attitude is seen as inadequate, and they no longer receive the respect the public had accorded them in the past. Civil servants are no longer viewed as impartial servants of the people discharging their responsibilities with efficiency, helpfulness and effectiveness.

Any attempt to change poor work attitudes and low staff productivity will be met with difficulties because these behaviours are institutionalised (Armstrong, 1999). For such change to be successful, it must be implemented as part of a continuous process change management (Johnson & Scholes, 2002; Johnson et al., 2017). The employee must own the change management process (Bernick, 2001) (Fig. 4.1).

Climate change has negatively impacted all aspects of life in African LDCs (Chimayo et al., 2011; IPCC, 2014a, b). Therefore, necessary technical and financial resources must be provided to leading organisations to develop appropriate climate change adaptation programmes to enhance the capacity of youth and their communities to adapt to climate change. This requires the injection of financial resources to drive the change process.

Burnes (2000) states that one of the barriers to the change process in organisations is the need to gain commitment from those who manage the organisation and the change subjects. Therefore, any consideration for applying the cultural web models in NAYCOM and other public sector agencies will depend on management's commitment to developing strategies that allow staff to buy into the change process (Johnson & Scholes, 2002; Johnson et al., 2017). Below is an illustration of a proposed sustainable culture (Fig. 4.2).

An Organisational Culture to Support a Sustainable, Positive and Dynamic NAYCOM Moving Forward

Stories

The future stories should concern celebrating the work and achievements of successful teams and demonstrating good ethics and values in achieving sustainable development goals locally and nationally. NAYCOM and the civil service should be promoted as a fun place to work by creating space for innovations and creativity in departmental and inter–/intradepartmental teams. The work and achievements of individual heroes and stalwarts in the civil service should be recognized. Such recognition must be done in the context of their effort to provide leadership, mentorship, positive attitudes, guidance and support to the teams to accomplish their various responsibilities and even go beyond daily requirements to seek opportunities that will benefit society. These practices should lead to building resilience and empowerment.

Routine and Rituals

Part of the ritual within the new NAYCOM/civil service should concern promotional campaigns to convey the new values and principles to be developed. The approach should consist of having continuous consultations with customers/clients to redefine the new customer–/client-focused model and feedback on how to engage people in providing quality services. The new model matrices should support information flow horizontally, vertically, bottom-up and top-down (Matland, 1995; OECD, 2013). Special dates and milestones should be identified and celebrated during the implementation process to bolster morale, increase transparency, accountability and trust and maintain a shared vision.

Symbols

Efforts should be made to create the enabling environment to foster informal interactions among all staff at all levels of the organisation where employees feel valued for their contributions to policy goals achievements. Team members should see senior managers' offices as

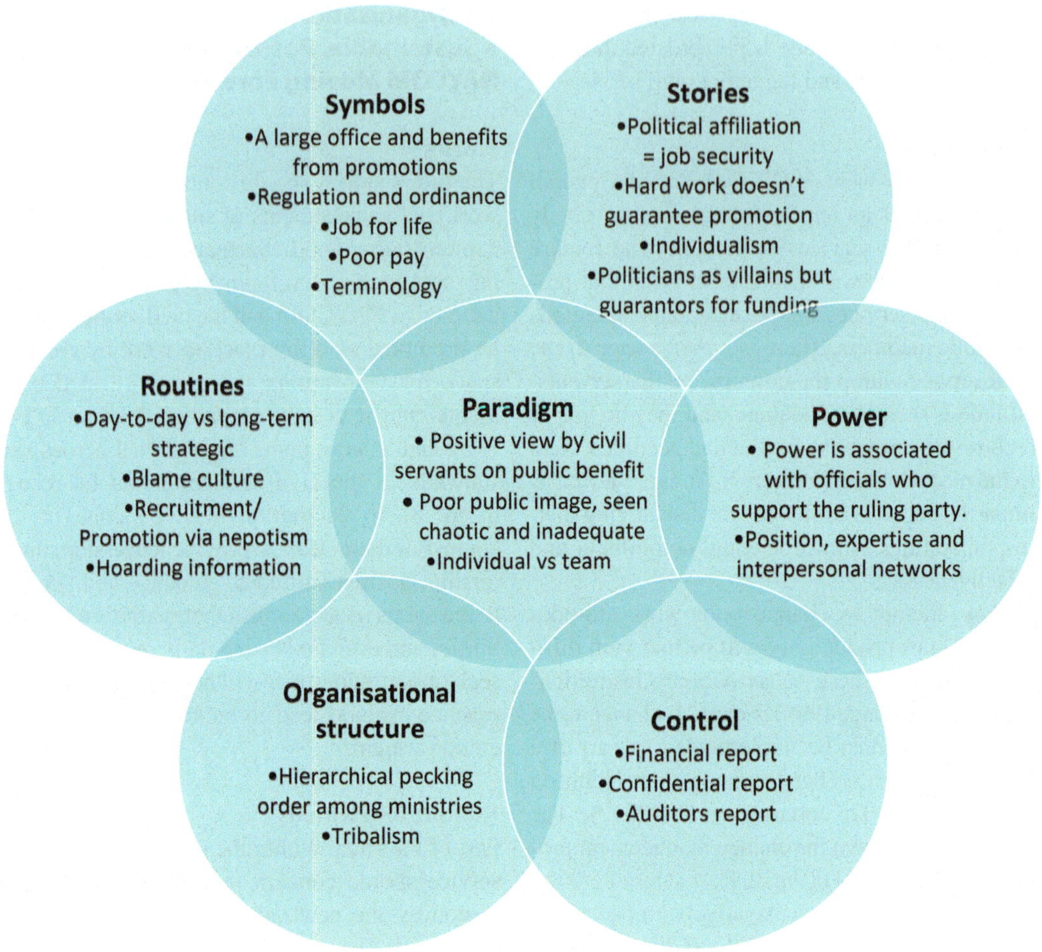

Fig. 4.1 The current culture of NAYCOM, the Government of Sierra Leone. (Adapted from Johnson & Scholes, 2002)

belonging and available to support work. It will be necessary to develop a new logo, indicating the new NAYCOM/public service's values, belief and paradigm.

Power Structure
The new power structure will value officials and teams who perform credibly on developing and fostering customer–/client-centred management practice. The model will encourage interpersonal networking to support information flow both vertically and horizontally and value officials who voluntarily share their expertise and competencies to benefit their team and the organisation.

Organisational Structure
The new NAYCOM will be less hierarchical and possess a flatter organisational structure. The structure would be more flexible in its approach and style and more responsive to the needs and aspirations of youths. It will be more conducive to incorporating partnership and a key stakeholder in its development role.

Control
Although financial and auditor reports will remain essential elements of control, these will be viewed from their effectiveness to achieve outcomes and proposed impacts. The Performance

Fig. 4.2 The future culture of NAYCOM, the Government of Sierra Leone. (Adapted from Johnson & Scholes, 2002)

Management and Development System and its related support systems (job classification review, HR Policy/strategy, etc.) will be instituted to enhance staff performance and provides equitable compensation for work performed. Electronic M&E systems will be implemented to provide adequate feedback to management on project/programme performance.

Paradigm

The new paradigm of the NAYCOM/civil service will develop a vibrant, dynamic production-oriented public service. It will implement a customer–/client-focused approach, building and strengthening teams to deliver quality service to the youth and their communities. Youth will see the NAYCOM/civil service as an agency that belongs to them.

The change process will involve changing employees' behaviours in changing cultural values to enhance sustainable outcomes, focusing on the organisation's paradigm. To achieve a shift, managers would implement education awareness initiatives, such as consultations with service users and providers, employees training and personnel development. Such information and scanning systems will keep managers and technocrats up to date with scientific and competitive research in the field (Johnson & Scholes, 2002; Sen, 1999; Johnson et al., 2017).

NAYCOM Information Technology/Information System (IT/IS) Operations

Analysis of Current IT/IS Application Portfolio

Senior managers view information and communication technology (ICT) as an essential component of NAYCOM and the broader public service to achieve their strategic goals. However, they do not fully understand the strategic contributions that ICT systems can play in supporting the achievement of policy goals. IT has not been a significant consideration in the strategic planning framework of NAYCOM. Unawareness of IT's strategic role in achieving policy goals and outcomes is evident in NAYCOM and other public service agencies in Sierra Leone (Earl, 2001). The ICT training offered to officials is basic, specifically on enhancing task at the operational levels.

The NAYCOM invested in the computerisation of some of its internal operation and the District Youth Centres (Ward & Elvin, 1999). NAYCOM has two areas of difficulties relating to IT's strategic use: (1) although its IT/IS systems are reasonable, they are not integrated nor compatible with each other; (2) although there have been significant efforts by NAYCOM to invest in appropriate technology at national and district levels, according to the Current Application Portfolio Matrix, such investment is not commensurate with realising policy outcomes (Fig. 4.3).

The operational and support application enables the organisation to avoid disadvantages in its business environment. NAYCOM has databases that can be readily accessed, including youth entrepreneurs who participated in business training at national and district levels and youth development consultants. Although efforts are made to computerise its accounting system and link it to the Ministry of Finance, the African Development Bank and UNDP IT networks to facilitate smoother implementation of project management processes, NAYCOM still maintains a manual bookkeeping ledger system.

NAYCOM PIU uses computers to prepare reports, including project management reports and M&E, and project design. Other applications and databases provide a supporting function. It has a reliable Internet system and website, mainly used for sharing and disseminating information. However, due to the unreliable nature of the Internet in Sierra Leone (it usually malfunctions for a considerable length of time), NAYCOM IT/IS systems are not performing at the strategic and high potential application levels. Although the senior management team (SMT) discussed IT's strategic use, these discussions do not materialise in its strategic application (Fig. 4.4).

The accounting system used by the NAYCOM is inefficient, ineffective, cumbersome and alien. It provides low business value and low technical fit to the organisation. Employees record database, payroll applications, etc. are of high technical fit to the organisation but are of low business value. These applications should be reassessed to decide how they could produce higher business value and support strategic decision-making and competitive advantage (Cavaye & Cragg, 1993). The Internet can be of high business value to the Commission but is not fully utilised. The Commission needs to look at more productive and efficient ways of reengineering the use of the Internet. The MS software has high business value and high technical fit; therefore, the Commission should retain its application and explore new ways of using them (Fig. 4.5).

NAYCOM needs to upgrade the IT desk and officers to a high-level in-house IT expert/consultant to professionally network its computers and integrate its divisional/district systems into one data model that supports data automated Web-Based Delivery System and Enterprise Resource Programme (ERP). Such systems would provide access to suppliers, DYC, partner institutions and other institutions supported by the organisation. This will allow for back-and-forth communication between the Commission, DYCs and partner institutions, allowing them to request supplies from the Commission or directly from suppliers. Provision training will be provided to partner institutions (tertiary, TEVT and BDS centres) via

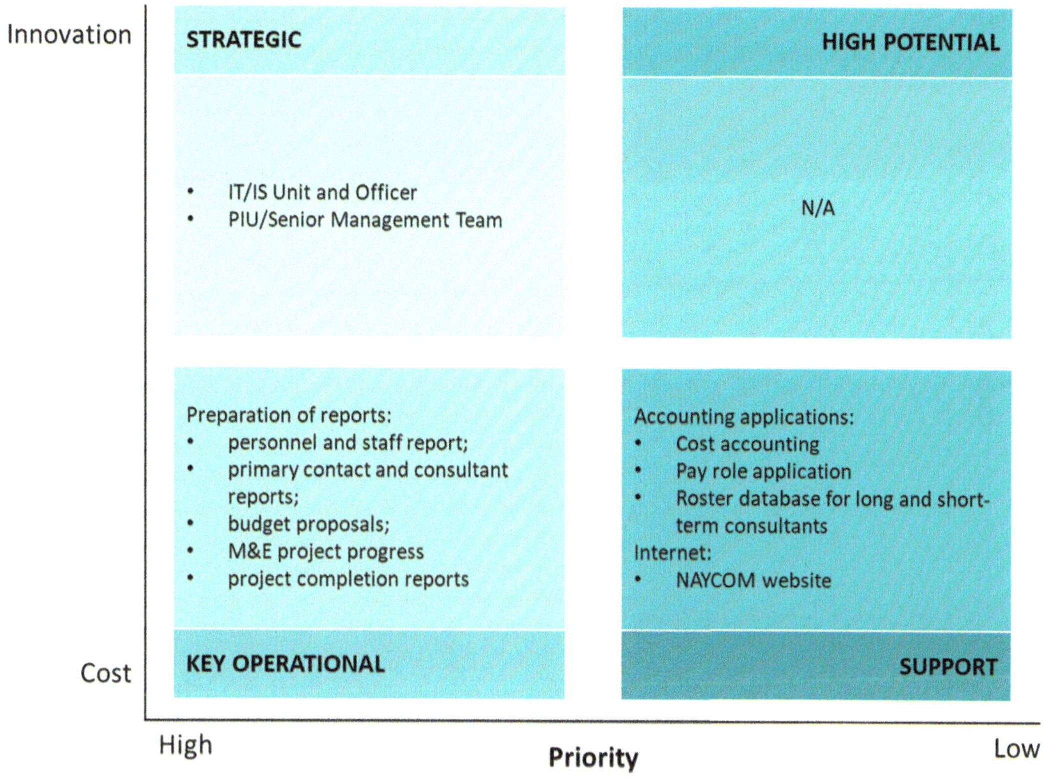

Fig. 4.3 Matrix of the current IT/IS model of NAYCOM. (Adapted from Ward & Elvin, 1999)

the Internet and access to partner institutions to NAYCOM electronic library and databases. These mechanisms will enhance existing systems to accommodate the new strategic roles of NAYCOM (Fergerson, 2012; Hickman, 2002; Thuraya, 2015; O'Brian, 2000).

A closer analysis of the Future Application Portfolio shows that the Commission will emphasise the Matrix's strategic and high potential components to achieve its strategic goals. In the high-potential area, the Commission will develop and test database applications that support the development of new service delivery approaches such as providing instructions to tertiary partner institutions, TVET and BDS centres and the use of e-learning via the Internet. An essential strategic innovation of the Commission will be to develop a corporate-wide IT system designed to provide 'Idea space' to support the generation of innovative youth development and entrepreneurship projects. The IT system will constitute an online marketplace for the development of radical ideas in education and enterprise development; therefore, anyone visiting the site can build on the ideas submitted by young people. Through crowdfunding and other funding innovation, the Commission should mobilise financial resources from donor agencies and local financial institutions to provide funding annually to the winning teams to develop their ideas into tangible projects. This initiative will provide opportunities for the NAYCOM to re-enforce its new culture through building on the highly efficient, electronic market for ideas, capital and talent promotion. Therefore, it will no longer be the knowledge management functions that constitute ITC's leading edge but the innovative marketplace (Hamel, 2001).

Assessment of the Use of the Flow Process Chart and Critical Path

Unemployment is exceedingly high among graduate youth in Sierra Leone. Although the govern-

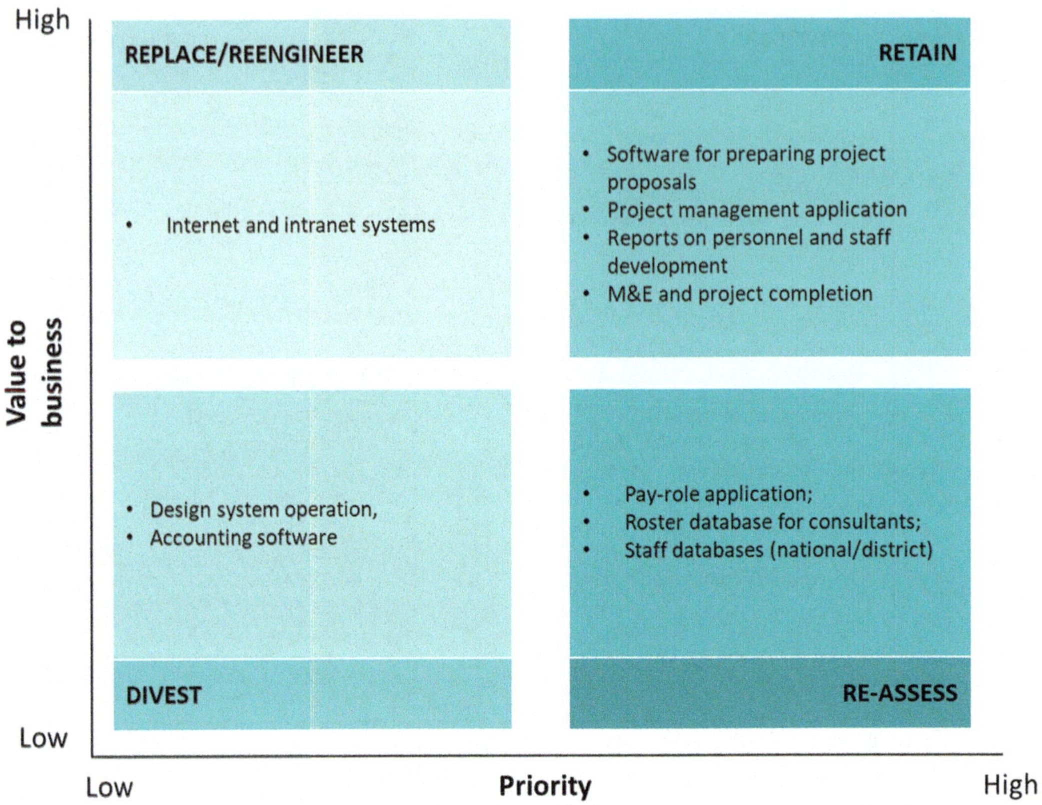

Fig. 4.4 Technical fitness of the system. (Adapted from Ward & Elvin, 1999)

ment ensures that places are available at tertiary and TVET institutions for young people to access education, the quality of the education offered by the institutions warrants urgent improvement. Research shows a mismatch between the quality of the training provided by education and training institutions and those required by the market (Carter, 2008, ILO, 2010; UNDP, 2015). The Ministry of Higher Education, in collaboration with the National Youth Commission, are in the process of implementing a comprehensive programme funded by the GIZ and the World Bank to enhance the quality of curricula, constructing quality school places and developing the competency of instructors through Training of Trainers and upgrading of tools and equipment to foster technical acceptance of TVET (Dilworth, 1996).

The consortium spent the first 2 months (September–October 2018) of the YEEP pilot project working with senior officials of the Commission to chart the YEEP flow from formulation to implementation. The Consortium team assessed the projects implemented by the government and realised that some projects with a timeframe of 3 years (36 months) were running for 5 and 6 years (60–72 months) and were not successfully implemented. This period almost equates to the same period of a government's entire term in office (60 months). Therefore, this project development model will not be attractive to any government, primarily when successful completion of such projects will provide opportunities for the government to be re-elected into office. The consortium team leader began work on the development of YEEP Gantt and PERT charts to track the implementation of the project.

The consortium concluded that the difficulties in using the traditional project management models are compounded by the complexities in coordinating the number of government ministries

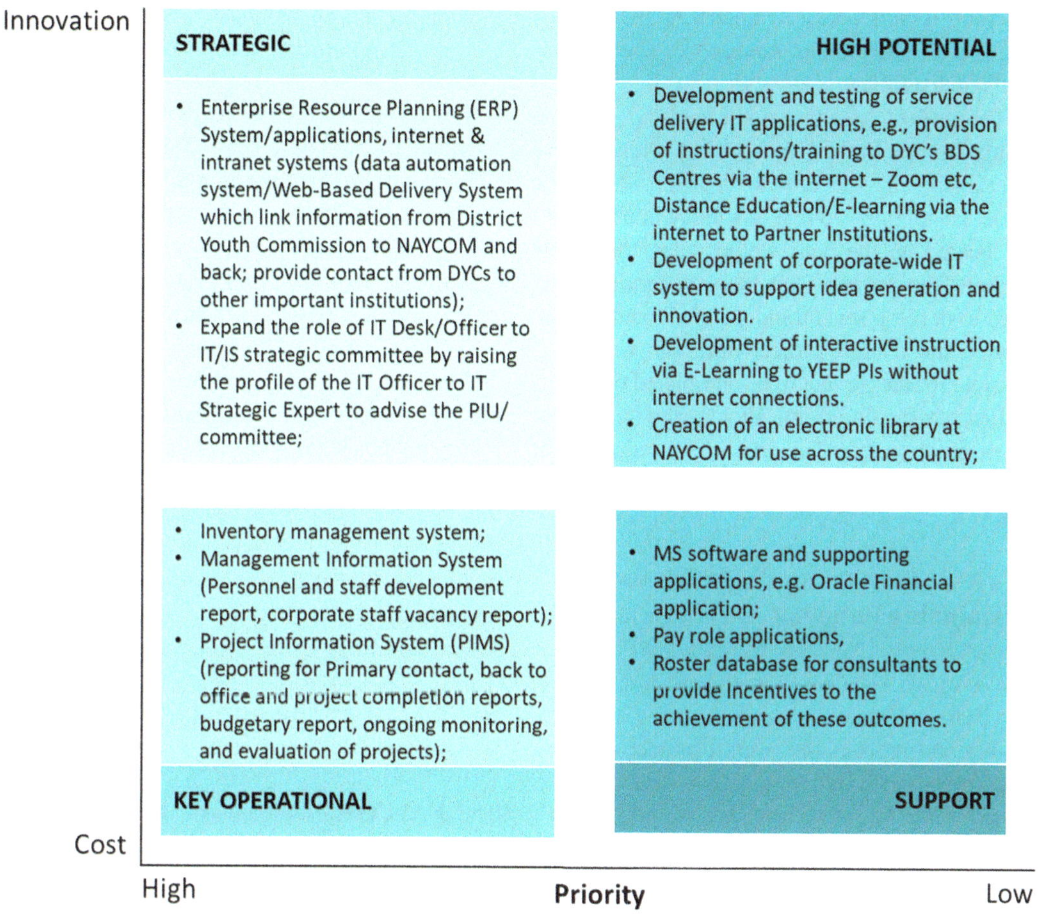

Fig. 4.5 Matrix to support strategic investment in IT; matrix of the future NAYCOM IT/IS model. (Adapted from Ward & Elvin, 1999)

and departments involved in creating employment for the youth sector. NAYCOM does not possess an effective project management system, neither does it have the capacity and capabilities to effectively manage all the agencies involved in the delivery of employment for youth in Sierra Leone. It must work in partnership with a range of ministries and institutions to ensure the project's effective delivery. Coordinating project across the civil service is overly complex and cumbersome. These governmental departments operate like independent states within a national state system, with their own top-down management and administrative and operational mechanisms (Matland, 1995; OECD, 2013).

Designing the New Sustainable Pathway

The critical path is defined as a chain of a sequence of activity beginning and ending after a project (Slack et al., 1998). It allows one to identify the essential activities in a project and calculate their duration. The project lead worked with the consortium and NAYCOM to identify the various independent's path in implementing the projects and devise strategies that can be performed simultaneously along the multiple stages of the project. He then identified the path through the network, which had the most extended sequence of activity in the project, and calculated the time it will take to implement the various stages and deliverables for each of the components' deliverables.

The project lead then identified the decision points in the life of the project cycle where assessment and the critical decision should be made. He recognised that any delay in any activity located on the path would delay the whole project. The project lead then manipulated the chart to develop several independent pathways to shorten the overall project duration, making it possible for the project to be completed 6 months earlier than the timeframe allocated to implement the original project. This indicates a reduction of 25% of the time taken to implement the project.

The Benefit of Designing the New Sustainable Pathway

The critical sustainable pathway and process flowchart enhances the project's efficiency and quality and reduces the overall project management process's duration and cost. Tools such as these:

- Provide an opportunity to better manage the project 'scope' to determine goals and work to be done.
- Assist with effective management of the human resources capacity of the project.
- Aid communication among manager and employees by enabling managers to provide adequate information to staff to allow them to perform efficiently by adding value to the different stages and deliverables on the project (Dilworth, 1996).
- Ensure the project is efficiently managed to enable completion according to the designated timeframe.
- Provide more significant benefits to customers and government due to shorter, effective and efficient implementation process.
- Assist the manager and consortium in keeping projects within budget. One must always bear in mind that project 'overrun' and delays can increase the project's cost.
- Support the development and application of project monitoring and evaluation systems/tools. The M&E manager at NAYCOM was highly effective in terms of the support provided to the consortium team to ensure the successful implementation of the project.
- Provide an opportunity to apply business process reengineering (BPR) by implementing innovative strategies and value the inputs and commitment of partner institutions.

The new sustainable pathway model provides the government with opportunities to achieve its policy goals and political mandates with greater efficiency and precision. The shorter duration will provide more significant possibilities to the government through NAYCOM to deliver better quality services at greater speed and efficiency to the public.

Reflection on the Application of the Assessment Tool

The NAYCOM is organised along with traditional functions (divisional lines) rather than along customer-focused processes. Dilworth (1996) argues that the process operates across functional lines and may involve many departments, making it essential to consider work from a process perspective. Therefore, there is an urgent need for rethinking, restructuring and redesigning the processes of NAYCOM to achieve performance improvements, quality of service and speed of delivery (Slack et al., 1998; UNDP, 2015).

The operations of NAYCOM are in different locations (headquarters in Freetown and District Youth Commission at district levels) scattered across the country with quite a considerable distance between them. This situation has profound implications for the attainment of policy outcomes as it provides undue stress on the management to coordinate its activities. It is further compounded by the absence of appropriate technologies and finances to support the coordination of these entities. Managers do not fully comprehend the vital use of IT at the strategic level to enhance the delivery of quality services. Therefore, the Commission must implement tools to assess the benefits of ICT to determine whether the organisation's strategic information system (SIS) investment is providing the anticipated results (Cavaye & Cragg, 1993). The Commission has a low capacity relating to ICT, project management and the implementation of other programmes. The Commission needs to put measures in place to enhance the capacity for e-learning within the organisation. Such actions are critical to determining the Commission's ability to deal with COVID-19 and Ebola threats in the future that can endanger the organisation's sustainability in the future (Earl, 2001; Bhalalusesa et al., 2013; Thuraya, 2015). The innovative e-learning strategies implemented by partner institutions during the COVID-19 pandemic enabled the consortium and partner agencies to continue to provide training to youth during the COVID-19 lockdown. These innovative strategies provided opportunities for the partner institutions to substantially increase the number of young people benefitting from the programmes.

Another critical issue affecting the Commission and other agents of the public service in Sierra Leone is the extent of political influence on policy outcomes and deliverables. Staff believe that the institution lacks sincerity because it fails to adhere to functional public service performance systems. Workers view the selection and promotion of senior officials in the public service as directly related to their political and tribal affiliation and not necessarily on their expertise and performance. In turn, they would appoint, promote and discipline officials according to that affiliation's policy and philosophy.

The negative impact of this belief has severe implications for the quality of performance and policy outcomes of the Commission and the success of any change programme instituted by this agency. Civil servants in the lower scales are convinced that merits for promotions are not resulting from effective, efficient and creative work but depend on personal networks and the nature of their political and tribal connections. It is important to note that staff believe that it is more rewarding to spend their time on social, tribal and political networking than on actual work on the Commission's projects. Senior managers at the Commission also identified tribal and political biases as a significant cause of poor quality and low public sector results.

It is also important to note that the Commission is not oriented along customer-focused lines. Staff members are mainly concerned with fulfilling the requirement stipulated in their narrow job descriptions rather than perceiving their job as adding value and contributing to addressing the needs of Sierra Leone's youth.

The staff used political affiliation, tribal and personnel networks and other patronages to ensure job security, promotion and training. Currently, the Commission lacks the capacity and financial resources to implement effective and efficient change programmes to enhance policy outcomes, efficiency and delivery. The low awareness of IT compounds this holistic use and the lack of adequate skills to strategically utilise ICT to improve the quality of service delivered to youth.

Therefore, the Commission must review its current policy management implementation process and explore a new policy implementation model. The Commission should examine reengineering for policy management and implementation to enhance service quality and shorten the timeframe for implementing youth development policies and programmes.

The use of the ethnomethodological and participatory approaches enables the consortium to gain employees' trust, insights and confidence. If the Commission is considering implementing the change management model in the future, in that case, it is crucial to hire persons whom the employer perceived as trustworthy, fair and having the credibility to drive the change process (Simmons, 2021). It is also essential to note that although the consortium developed a comprehensive cultural web and IS/IT systems/models to support the effective implementation of change management machinery, due to the COVID pandemic, the Commission was unable to implement these strategies.

Section 2: Factors Utilised to Drive Successful Implementation of the Project

The second section of the chapter discusses the challenges, methodologies and processes the consortium utilises to implement the project deliverables successfully. These factors are discussed along the lines of the deliverables of the components successfully delivered by the consortium.

Component 1

The Skills Development for Employment covers seven tasks and deliverables. These include the following milestones (Table 4.1).

Curriculum Modular Development

The skills development curricula were developed with inputs from experts and institutions and sent to NCTVA for review and approval. The curricula were designed to provide the appropriate guide and pedagogy for training students in partner institutions to enhance their employment skills to reduce the skills gaps in the labour market and promote youth employability (Williams, 2013; UNDP, 2015; UNESCO, 2014, 2015; ILO, 2010). The curricula contain training guidelines, methodology and supporting materials for trainers/tutors to execute the specific skills training. The content is sector-specific, and the goal is to enable lecturers and supervisors to facilitate effective learning with positive results. The focus of the new curricula is to prepare youth to

Table 4.1 Component 1 results framework

Outputs	Status
Developed PPP framework	The PPP framework was completed and approved by NAYCOM (Chap. 3)
Reviewed and updated course content and training methodologies for the three priority skills and trades in selected TVET institutions	Practical training in Sustainable Construction (Carpentry, Mmasonry and Aluminium Ffabrication), Sustainable Agriculture (emphasis on Aquaculture) and Sustainable Hospitalities (Hotel Management and Tourism) was all reviewed. New Climate Change Education for Sustainable Development curricula in the three priority areas was developed and approved by NCTVA. Training in the priority areas was conducted in the various partner tertiary and TVET institutions. Students were sent on job placements to enhance their practical experiences and knowledge in a different environment
Entrepreneurship curriculum for the selected TVETs	The NCTVA approved the curricula and participated in the graduation and certification of students under this component. The curricula utilised the Climate Change Education for Sustainable Development model. NCTVA's Director was a signatory on the certificates, indicating that the students completed all the requirements leading up to the course
Training of trainers for *20* TVET instructors for the priority skills and trades and entrepreneurship training	Sixty-three (63) TVET instructors from tertiary and TVET institutions were trained in new content and methodologies for delivering training to youth in TVET employment skills. TVET institutions used the training contents as a guide during training on priority and supervisory skills. Training tools were supplied to each partner institution by NAYCOM/AfDB under component 1 of the project to facilitate practical sessions for effective teaching and learning, installed and are currently used by these institutions
Training a minimum of 100 youths in supervisory skills	The supervisory training was implemented at the three PIs. A total of 104 participants (M, 61; F, 43) from the different partner institutions were trained in supervisory management skills. The participants improved their teamwork skills, became more result-oriented, maintained the focus of their goals and desired to improvise and take the initiative
Training 150 youth in the selected priority skills	Overall, 325 youth (M, 181; F, 144) were trained in three priority skills from three PIs, including GTI, MMCET and Njala University. An MOU was drafted between GTI and MIRO for the private sector to provide resources to assist the government's TVET institutions in delivering employment training programmes to youth. Discussions and meetings were held with hotels and travel and tourist agencies (Tourism is Life, Visit Sierra Leone), an agro-fishing company in Bo and construction companies to create partnerships and internship opportunities for students

enhance their resilience to adapt to climate change and other development calamities such Ebola epidemic and COVID-19 pandemic.

Relevant skills development is critical to youth employment and job creation in West Africa. The consortium embarked on this component with a study of the status of youth employment in Sierra Leone using a range of research methodology and tools. The study discovered that youth in Sierra Leone have inappropriate skills to enhance their employability (ILO, 2010; Carter, 2008) due to the non-availability of appropriate tools and training received from TVET institutions. TVET institutions have the infrastructure to carry out the technical training; however, they do not have curricula, modern equipment and tools to train students in the respective skills offered by their institutions.

Training of Supervisors and Mentors

Three 4-day training sessions were implemented for PIs, facilitated by the private sector and YEEP team. PI staff were trained by selected private sector instructors and experts on the new curricula using appropriate methodology and pedagogy with a predominance in practical sessions. Completion of the training resulted in the certification of participants by NAYCOM.

One hundred and four (104) youths (M, 61; F, 43) were trained in management and supervisory skills from the three institutions (Njala University-Mokonde Campus, MMCET & GTI). The training enhanced their mentorship and supervisory skills. Implementing the PPP framework is pivotal in curriculum training as it enables the private sector to have access to professionally trained students in various skills areas by training the trainers.

Employment and Employability

Once students received the requisite training and complete internships or job placements, they are equipped with the skills and experiences needed by employers in their chosen fields. YEEP intends to enhance youth employability in Sierra Leone through training on 'skills for development' in TVET institutions. PIs and private companies established PPPs during the pilot project, which have increased due to the growing awareness of the benefits for both sides. A MOU has been signed between MIRO, a sustainable timber company involved in agroforestry and furniture production, and the Government Technical Institute. MIRO shared knowledge on innovations in furniture and lumber production, technical and material support, tools and equipment and job placement opportunities with the students and staff of GTI.

A series of meetings were held with hotels and restaurants and travel and tourist agencies (Tourism is Life, Visit Sierra Leone, etc.) to secure internship opportunities for students from MMCET. Thirty-four (34) students from the Aquaculture and Fisheries Management Department, School of Natural Resource Management, Njala University, were placed on an internship at an agro-fish farm in Bo, Njala University Fish Farm Mokonde and Mano Daise in Moyamba District. Other private sector players were contacted to participate in the internship/placement programmes.

Impact of the COVID-19 Pandemic

The COVID-19 pandemic has severely affected youth trained in Sustainable Hospitalities (Hotel Management and Tourism). They were unable to carry out their internship placements with hotels and restaurants across Sierra Leone. The hotel industry is severely affected by the pandemic. Approximately 80% of hotels and restaurants are either closed or partially functioning since the country went into lockdown on 21 March 2020 (UNDP, 2020). Over 75% of students expected to gain employment after participating in internship and apprenticeship activities in the tourism and building/construction industries were not put on placements. However, due to the entrepreneurship training provided, many graduates completing their training collaborated with peers to set up businesses.

Component 2 (Table 4.2)

Research on Career Guidance and Job Readiness Skills

The research was conducted using appropriate qualitative research methodology. The outcomes were that graduates mistakenly choose courses in the university, colleges and TVET that do not match the expectations of the job markets; hence they were unable to gain employment after graduation. This is due to the absence of appropriately trained counsellors and career guidance officials in these institutions. Career Guidance and Job Readiness (component 2 of the YEEP project) was appropriate to help address these growing issues affecting young graduates from tertiary institutions in Sierra Leone.

Table 4.2 Component 2 results framework

Outputs	Status
Training/ operational manuals for career guidance and job readiness program	Approved curricular and training manuals were finalised and used throughout the Career Guidance and Job Readiness training for students in universities and colleges in Sierra Leone
TOT for 30 counsellors/ trainers	60 counsellors from universities, colleges and tertiary institutions completed training in TOTs. Continued use of the online platform by PIs to enhance a sustained learning environment
Training/ counselling 2400 youth in Career Guidance and Job Readiness	The consortium conducted effective monitoring of the quality of all training provided by PIs. CGJR skills training was successfully completed. Graduation ceremonies were organised for each partner institution, and certificates of completion on the Career Guidance and Job Readiness course were issued to students by NAYCOM. All stakeholders (NCTVA, NAYCOM, PIs) and the contractors signed and authenticated the certificates, making them acceptable by employers in Sierra Leone and the other West African States

The Career Guidance and Job Readiness Skills programme is accessible to students in the participating institutions, including SOS, MMCET – Goderich Campus, EBK, Eastern Polytechnic, GTI and Njala University (Bo and Mokonde Campuses). Sixty (60) counsellors/university lecturers were trained on Career Guidance and Job Readiness modules to deliver training to the targeted group of 2400 students. During the roll-out at the institutional level, 6200 students successfully completed the training. The increase in enrolment is a result of (a) the high interest expressed by students which is due to emphasis on training students to acquire soft job skills, (b) the e-learning programme introduced by the institutions as a strategy to engage students on their academic work during the COVID-19 pandemic which eluded the physical restrictions on the number of students that could participate, and (c) the use of working groups, WhatsApp and Facebook to deliver course content which built networks, enhanced friendship and made the training more attractive and accessible to young people.

Curriculum and Modular Development

Career Guidance and Job Readiness experts collaborated with counsellors across the country to develop the Career Guidance and Job Readiness curriculum and training manuals. These were designed based on the UNESCO Climate Change Education for Sustainable Development model (William, 2013; UNESCO, 2014). The documents were validated at a national workshop implemented by the consortium. The curriculum holds the training methodology and topics to be covered during the actual training of students, which was successfully completed in all institutions.

Training of Counsellors and Instructors

Training of Trainers sessions, guided by the curriculum developed for Career Guidance and Job Readiness programmes, were conducted for PI lecturers, counsellors and staff. Two 4-day training workshops were held to train counsellors delivering the course in Bo and Freetown in the new pedagogy, methodologies and content relating to enhancing skills and competencies. A total of sixty (60) counsellors were trained on Career Guidance and Job Readiness modules; this is two times the number of counsellors anticipated by the project.

Training of Young People

Students were recruited and trained by counsellors, mentors and lecturers in Career Guidance and Job Readiness skills. The 9-month training programme was student-centred and practical in orientation, imparting career development soft skills such as time management, job application and the impact of climate change and development calamities on emerging careers. Online platforms were created to foster interaction and communication among youth, counsellors and lecturers during the national lockdown when universities and colleges were closed during the COVID-19 pandemic when social distancing and other preventative measures were instituted. It was wonderful that students' enthusiasm remained extremely high during this period.

Component 3 (Table 4.3)

GEP TOT and Reorientation

Following the approval of the GEP training manual and operational instructions by the PIU of NAYCOM, the entrepreneurship TOT was designed and implemented to target 16 trainers from selected PIs and 4 representatives from NAYCOM. The GEP training was executed in three selected PIs based in Freetown, Kenema and Makeni, respectively. The GEP implementation further created the opportunity to build a network of mentors capable of supporting and implementing the Graduate Entrepreneurship Programme in Sierra Leone. A GEP platform was developed to facilitate such collaboration; graduate entrepreneurs could interact with mentors and peers to share information on resource mobilisation, knowledge and best practices. The platform is web-based, allowing youth entrepreneurs to partner with funders and other entrepreneurs within and outside Sierra Leone. It is early to assess the full capacity of the web-based platform; however, it is hoped that the model's performance will produce the result anticipated.

Table 4.3 Component 3 results framework

Outputs	Status
Business service centre (BSCs) staff orientation	Twelve (12) BSC staff and four (4) NAYCOM backstopping staff were trained in implementing the Graduate Entrepreneurship Programme in their respective institutions. Capacity-building sessions were regularly held for facilitators throughout the YEEP programme
Prepared and customised training manuals and operational instructions for the entrepreneurship programme	A training manual for the GEP was developed and delivered to each partner institution. The training contents and proposed methodologies were practical-oriented and were found to be beneficial to graduate entrepreneurs
Installed entrepreneurship programme including training of 60 youths	105 (M, 63; F, 42) graduated successfully from the GEF programmes. These young people were supported and linked to external mentors and business coaches for onward coaching and partnership
Established local and international networks to provide business counselling and mentoring to youth	BSCs have been linked to the development of the digital platform. The platform is linked to the NAYCOM/GEP website: GEP-NAYCOM.b4dev.net
Designed and established a youth entrepreneurship platform for networking and exchange of ideas	The GEP digital platform was developed and launched. The capacity building of pivotal NAYCOM staff was conducted to manage and control contents published on the platform: GEP-NAYCOM.b4dev.net

GEP Training of Youth

A total of 105 young entrepreneurs were trained instead of the original target of 60. Achieving the GEP goal requires graduate entrepreneurs to go through a process from the ideation stage to the acceleration stage, enabling them to develop bankable business ideas and project proposals. The ideation stage focused on equipping the youth entrepreneurs with business strategy using the

Business Model Canvas as a principal orientation and training tool. Coaching also formed part of this process as it enabled entrepreneurs and mentors to discuss critical issues in their product development. During the second stage of the training process, the incubation stage, three thematic areas were emphasised for attention, including market research, operations and human resource and financial modelling. Here, the graduate entrepreneurs would conduct intensive market research to develop a marketing strategy, develop economic models and test their businesses' viability through financial management tools available during the training programme.

The acceleration stage enhances the skills of the young entrepreneurs in managerial competence, operational ability and growth potential of their entrepreneurial ventures, building business partnerships and attracting investors and bankers through pitch development and presentation. Due to COVID-19 restrictions, this activity was done within the partner institutions rather than at the national level. At this stage, effective coaching and mentorship support are critical to helping the youth entrepreneur to mature. The young entrepreneurs were assigned to mentors and business coaches with speciality in their respective businesses. The bankable business plans were developed and shared with their respective coaches, mentors and trainers.

Reflection on YEEP

Like any other development project, YEEP is faced with its own set of development challenges, gaps and strengths. In this section, the weaknesses and strengths from the perspective of the stakeholders will be explored. This evaluation was implemented during the peak of the COVID-19 pandemic, 6 months before the project ended in October 2020.

Reflection on the Weaknesses of the YEEP

A senior government officer highlighted the weakness of the YEESP as:

> The major challenge facing the project is with the consortium, the implementing body of the project. Gaps exist between the team operating in Freetown, Sierra Leone, and the body that manages the project in Milan, Italy. While the team in Freetown was robust and ready to move the project forward in terms of its implementation, there was evidence of delay in the implementation of actions to move the project forward between management and the team in the field in Freetown, resulting in the delay of the implementation of project activities. (Senior Government Officer – 1GN)

Mistrust was highlighted as another development challenge facing the project. Mistrust was highlighted as responsible for creating a further distraction to the implementation of the project. This was indicated by a senior officer of a partner institution:

> There is a lot of mistrust between the consortium and the YEEP project manager, which has distracted the project's overall success. This is an issue that needs to be addressed moving forward. (Senior officer – Partner Institutions (2GN))

Although the senior managers of PIs were concerned with issues of mistrust and lack of timely support to the project implementation team in Freetown from their counterparts in Milan, representatives of PIs were more concerned with the absence of tools and equipment on the ground to implement the practical aspects of the training courses in the partner institutions. Here is the view of the youth:

> The programme is plagued with a range of challenges, including a) lack of equipment, b) lack of industrial gears to facilitate the delivery of training, and c) poor financial support towards the implementation of the programme. (Youth participants – 3C1)

From all assessments, youth are excited about the YEEP. However, youth are also concerned with the lack of tools and equipment because the longer the delays in receiving the tools and equipment, they would lose out on the practical nature of the courses they are pursuing; the courses are assessed as 60% practical and 40% theory. The opinion of an instructor on the course is:

> There is nothing that I dislike about this course. I like everything. However, the lack of learning material and equipment has been creating prob-

lems for lecturers and instructors alike. You cannot teach without tool and equipment to demonstrate examples and or certain concepts. (Training Instructors – BC1)

Although the project achieved all its deliverables, there were weaknesses in the project implementation. Different stakeholders view the challenges/weaknesses from different perspectives. While the lack of support provided to the implementing team in Freetown by the consortium based in Italy was of concern to government officials in Sierra Leone, the delay of tools and equipment provided by NAYCOM was the main concern raised by senior officials of PIs and the youth participating in component 1 of the project. Tools and equipment were critical for PIs to implement the new sustainable curriculum for the 'Skills for Employment' programme. The tools and equipment were to arrive in August 2019 for the beginning of the academic school year in October 2019; however, they did not arrive at the PIs until June 2021. The tertiary institutions lost a considerable amount of time when the institutions were closed due to the COVID-19 lockdown.

Both challenges raised by the stakeholders are important and critical for the successful implementation of the project. The consortium team on the ground in Freetown needed financial resources to implement the various project activities associated with each of the project stages and deliverables. The lack of adequate resources available to the team in Freetown resulted in either postponement or cancellation of some critical project activities. This has created severe tensions and misunderstanding between the YEEP project managers, NAYCOM staff and the local consortium staff based at the project office. Tension was also high between the consortium and PI officials, where PIs, on many occasions, decided not to implement any training until they were paid for the work according to the accepted MOU. These issues were compounded by the unannounced monitoring visits of the NAYCOM PIU team, which contributed to inflaming the tensions between PIs, NAYCOM and the consortium team. NAYCOM PIU took a considerable amount of time to approve project stages, according to the consortium's management team; the delay contributed to the non-approval of the deliverables and further delays in the payment from the bank to the consortium's head office.

The UNDP (2015) report states that the challenge in the skills mismatch in the knowledge economy is due to the slow pace of technological innovation, compounded by the rapid growth in demand for new and higher skills, which are not forthcoming in most developing countries (Paschen, 2015; ILO, 2010). Therefore, it is critical for the latest tools and equipment to be provided to PIs to avoid such developmental pitfall; delays due to a lack of urgency or mistrust only exacerbate the workforce skills gaps, putting Sierra Leoneans at a disadvantage nationally and globally.

The mid-term report (MTR) was submitted 5 months before a response was given by the PIU of NAYCOM, which considerably contributed to the delay of the project implementation process. The commissioner, who was frustrated with the constant delays of the project timeline, agreed to take full responsibility for implementing the remaining project deliverables, working directly with the consortium project lead to implement the project. This decision did not sit well with the YEEP project manager and the members of the PIU. Although the consortium team and NAYCOM staff agreed joint visits during monitoring and evaluation activities to PIs (IGN), this did not happen until the end of the project. The visit provided opportunities for both NAYCOM and the consortium to address critical issues facing partner institutions and address mistrust issues that were confronting both agencies.

Although senior officials of PIs were given information on the project's status, this information remained with the management of PIs. They did not inform the officials responsible for delivering the training programmes, resulting in misinformation and mistrust towards the consortium, NAYCOM and PI senior management teams. The consortium team took a considerable amount of time travelling from one district to the other to relay appropriate information to PIs on the project and provide transparency and accountancy in the project processes.

Reflection on the Strengths of the YEEP

A senior government officer posits that the YEEP has several unique strengths which contributed to the high levels of success achieved. Other senior government officials support this view:

> The project is addressing the direct needs for employment through skills training of youth in the country. Several reports came out over the years stating that youth in Sierra Leone are severely unemployed. The project is addressing one of the severe development challenges/issues facing the country at hand. It is focused on addressing the issue of enhancing the employability of youth of Sierra Leone through entrepreneurship. (3 Senior Government Official – CN)
> The programme has unique strengths. It uses the MMCET infrastructure/training facilities and trainers to deliver the training programme to the youth. It has many students who are motivated and ready to learn as well as a cadre of lecturers who are full of enthusiasm and are ready to deliver their knowledge and skills to young people. (Senior PI Official – 3C1)

Another senior government official recognised the uniqueness of the strategy used by the consortium and noted that it provided mutual benefits to the local PIs, the YEEP-implementing consultancy and the youth of Sierra Leone:

> The strategy of outsourcing the implementation to local institutions is a novel innovation of the project. The capacity of the local institutions (PIs) involved in the implementation of the project is also being enhanced by the experts provided by the consortium. The enhancement of the capacity of officials and heads of PIs is provided through technical advice, mentoring & coaching, and training (ToTs) by the consortium expert. (Senior Official of Government – 2GN)

Youth enjoyed participating in the project. One youth states that the project works due to the unique partnership forged between the consortium and PIs:

> I like the project because it focuses on providing skills training to youth to improve our prospects for Employment. The training is working very well because there are lots of qualified and experienced instructors and counsellors available to deliver the training to students. (Youth participating in the YEEP – FC1)

Reflection on the Strengths of the Project

All stakeholders attested to the unique strengths of the project attributed to the high participation of students in the programmes. It was responsible for the high enthusiasm the lectures/counsellor and other stakeholders have for the skills training programme. One PI representative stated that the strategies employed by the consortium and the PIs, along with the lectures, created an enabling environment for youth to participate in the training courses. He stated that all the youth registered in the course attended all aspects of the programmes. This is not the norm in most training courses implemented in the institution in Sierra Leone.

A key strength highlighted by the senior government officials is that the components of the project were contracted to local institutions through the MOUs, and therefore responsibility and ownership were given to implement the project within their respective institutions and communities. This model displays unique attributes in the coordination of the project. It provides opportunities for the consortium and NAYCOM to enhance the capacity of local partner institutions involved in the implementation of the project. The consultative and participatory approaches implemented by the consortium contribute significantly to improving the governance, management and overall institutional capacity of these local institutions to come up with solutions to resolve the challenges facing the management and implementation of the project (Simmons, 2021; IUCN & WRI, 2014; Baxter & Jack, 2008). It also provided opportunities for the PIs to contribute to the realisation of national sustainable development goals by contributing directly to capacity development, youth development and employment creation (Government of Sierra Leone, 2019).

In coordination and networking, the project has involved a diverse range of partner institutions in youth entrepreneurship and employment across all development sectors in Sierra Leone.

There is a strong sense of synergy from different institutions learning from each other as they progressed in implementing the project; strong linkages with civil society organisations and NGOs working in the field were evident and supported by IGOs across the country. YEEP was expanded out of Freetown to include partnerships with credible institutions in other urban and rural centres providing incredible reach for the project and contributed to doubling and tripling the beneficiaries in the project's components.

The youth are pleased with the results of the project and expressed interest to participate in the project and its upscaling because the project focuses on addressing fundamental needs that are personal to them, i.e. improving the skills, knowledge and competencies of youth to find and/or create their own employment. To the youth, collaborating with lecturers in the various training courses has worked very well because both the lecturers and students were very motivated and committed to the programme. The skills component of the project focused on building the resilience and adaptive capacity of youth against climate change and other development calamities such as Ebola and COVID-19 through the development of climate-friendly and sustainability-focused curricula in the three skills areas. The strategies implemented by the consortium were essential to protect enterprises, workplaces and communities from the negative impacts of climate change and development calamities in the future (ILO, 2013; Nurse et al., 2014; Paschen, 2015; UNDP, 2015).

YEEP supports the adaptative options that are feasible for youth development ranging from training, capacity building, institutional building, social assistance and local infrastructure and technological development. YEEP places the impetus to increase sustainable change and support youth development in the partnership between the private and public sectors, but most notably with individuals within these organisations to champion and drive excellence. Institutions are actively acquiring new partnerships with private organisations and developing their infrastructure to support better skills development for their students. Njala University is currently building aquariums for its aquaculture programmes. Similarly, MMCET is building an e-booking facility for its sustainable hospitality programmes, and GTI is creating a fabrication facility to expand its sustainable construction training course.

Summary of the Chapter

When organisations focus on the strategic application of technology, utilising change through assessment of organisational culture and business process reengineering strategies, covered in this chapter, change management is boosted, and decision-making support is enhanced, resulting in increased efficiency in their operational mechanism and competitive advantage.

A strong culture within an organisation is a powerful lever to drive change (Deal and Kennedy 1982 in Burnes, 2000). Cultural theorists such as Hofstede believe that cultural influences play a critical role in the way employees behave inside and outside their organisation (Beardwell & Holden, 1997). Therefore, to create effective change in an organisation like the National Youth Commission, change managers need to conduct a thorough assessment of the organisation's current culture and make amendments to their cultural web to determine the type of organisation they are considering recreating for the future. A thorough assessment was implemented on the cultural web levers, including the routines and ritual, power structures, organisational structure, symbols, stories and paradigms (Johnson & Scholes, 2002; Johnson et al., 2017).

NAYCOM has a role culture, where implementing changes to enhance efficiency and effectiveness is met with challenges due to its inherent rigidity and culture of complacency (Handy, 1986). Through discussions with senior management and staff, the consortium proposed an alternative culture to enhance the efficiency and effectiveness of the organisation in line with the government strategy to enhance its human capacity to achieve global sustainable development goals (Government of Sierra leone, 2019). However, the consortium could not run a change management programme due to the outbreak of the COVID-19 pandemic. The pandemic meant the team needed to prioritise attaining the YEEP's deliverables rather than modifying the organisation's culture.

NAYCOM utilises IT/IS systems and the Internet to modernise its governance, management and operations mechanisms. The information technologies and information system (IT/IS) reviewed how information technology is used to support management to make strategic decisions to enhance the overall operation of the project. The outcome of the assessment showed that although NAYCOM established an IT desk staffed by an IT officer, an IT strategy to indicate the role ICT would play in assisting the organisation's management in achieving its policy goals is absent (Earl, 2001). This is compounded by the absence of programmes within the organisation to raise awareness on how IT can give the agency a competitive advantage. IT is primarily used to support low-level office operation. The consortium had developed a comprehensive programme regarding the investment needed to enhance the IT system and the future training required to ensure that ICT is utilised at the strategic level to improve the organisation's management and overall operation; however, this was not implemented for the same reason as the culture change programme.

Project drift is a serious consideration affecting the implementation of projects in developing countries. Many projects implemented by the donor agencies in Africa running behind their implementation schedule failed and/or abandoned project deliverables (Okereke, 2017). There is evidence where projects with a 3-year duration would drift to as much as 4 to 5 years. The consortium utilised technical programmes such as business process reengineering, new sustainable pathways and PERT plan applications to manipulate the pathways to reduce the timeframe for implementing the YEEP. Despite the outbreak of the COVID-19 pandemic, which forced the government to close all colleges and universities and caused a delay in funding and implementation of the project, the consortium successfully delivered all the project deliverable and tripled its target beneficiaries. The consortium achieved the above through the implementation of flexible and robust strategies consisting of (a) the transparency of the consortium to contact PIs and dispel misinformation and rebuild trust in the project, (b) the high calibre of training supported by the private sector to bolster student skills and entrepreneurship, (c) the use of technology to increase the number of students able to take part and be supported and (d) the provision of financial support to PI to enable them to implement the project. These strategies contributed to the high levels of enthusiasm among all stakeholders participating in the implementation of the project.

References

Armstrong, M. (1999). *How to be an even better manager* (5th ed.). Kogan Page Limited, Pentonville Road.

Baxter, P., & Jack, S. (2008). Qualitative case study methodology: Study design and implementation for novice researchers. *The Qualitative Report: McMaster University, West Hamilton, Ontario, Canada, 13*(4), 544–559.

Beardwell, I., & Holden, L. (1997). *Human resource management: A contemporary perspective* (2nd ed.). Pearson Education Limited, Prentice Hall Publishing.

Bernick, C. L. (2001). When your culture needs a makeover. *Harvard Business Review*, 53–61.

Bhalalusesa, R., Lukwaro, E. E., & Clemence, M. (2013). Challenges of using e-learning management systems faced by the academic staff in distance-based institutions from developing countries: A case study of the Open University of Tanzania. *Huria Journal of OUT, 14*, 89–110.

Burnes, B. (2000). *Managing change – A strategic approach to organizational dynamics* (3rd ed.). Pearson Education Limited, Prentice Hall Publishing.

Carter, R. (2008). *Caribbean youth: An integrated literature review*. Bridgetown.

Cavaye, A. L. M., & Cragg, P. M. (1993). Strategic information system research: A review and research framework. *Journal of Strategic Information System, 2*(2), 125–137.

Chima, O. O. (2017). Causes of failure and abandonment of projects and project deliverables in Africa. *PM World Journal, VI* (I). www.pmworldjournal.net

Chimayo, E., Okali, D., Kowero, G., & Larwanou, M. (2011). *Climate change and African forest and wildlife resources*. African Forest Forum, Nairobi, Kenya.

Dilworth, J. B. (1996). *Operations management* (2nd ed.). McGraw-Hill Publishing.

Earl, M. J. (2001). IT strategy in the new economy. In J. Pickford (Ed.), *Mastering management 2.0: Your*

single-source guide to becoming a master of management, financial times. Prentice Hall Publishing.
Fergerson Brian. (2012). Key Stages of Strategic Information System Planning (SISP) Methods and Alignment to Strategic Management Planning Concepts: University of Oregon Applied Information Management Program: Continuing Education 1277 University of Oregon Eugene, OR 97403-1277 (800) 824–2714.
Government of Sierra Leone. (2019). The Government of Sierra Leone's new Medium-term National Development Plan (MTNDP) 2019–2023, Freetown.
Hamel, G. (2001). *Leading the revolution*. Harvard Business School Press.
Handy, C. (1986). *Understanding organization* (4th ed.). Penguin Books.
Hickman, S. (2002). *Handout on Strategic Information System Planning (SISP) – Information technology and business, masters in business administration*. University of Greenwich.
ILO. (2010). *A skilled workforce for sustainable growth: A G20 Training Strategy*. International Labour Office, Geneva.
ILO. (2013). *Sustainable development, decent work and green jobs*. International Labour Conference, 102nd Session, International Labour Office, Geneva.
IPCC. (2014a). Climate Change (2014): Synthesis report. In *Contribution of Working Groups I, II and III to the fifth assessment report of the Intergovernmental Panel on Climate Change* (p. 151). IPCC.
IPCC. (2014b). Summary for policymakers. In C. B. Field, V. R. Barros, D. Dokken, K. J. Mach, M. D. Mastrandrea, T. E. Bilir, M. Chatterjee, K. L. Ebi, & Y.O. Estrada (Eds.), *Climate change impacts, adaptation and vulnerability* (Contributions of the Working Group II to the fifth assessment report). IPCC, Geneva.
IUCN & WRI. (2014). *A guide to Restoration Opportunities Assessment Methodology (ROAM): Assessing forest landscape restoration opportunities at the national and sub-national level* (Working paper (road-test edition)) (p. 125). IUCN.
Johnson, G., & Scholes, K. (2002). *Exploring corporate strategy* (6th ed.). Prentice-Hall.
Johnson, G., Scholes, K., & Whittington, R. (2017). *Exploring corporate strategy: Text & cases* (11th ed.). Pearson Education Limited.
Matland, R. (1995). Synthesising the implementation literature: The ambiguity-conflict model of policy implementation. *Journal of Public Administration Research and Theory, 5*(2), 145–174.
Nurse, L. A., McLean, R. F., Agard, J., Briguglio, L. P., Duvat-Magnan, V., Pelesikoti, N., Tompkins, E., & Webb, A. (2014). Small islands. In V. R. Barros, C. B. Field, D. J. Dokken, M. D. Mastrandrea, K. J. Mach, T. E. Bilir, M. Chatterjee, K. L. Ebi, Y. O. Estrada, R. C. Genova, B. Girma, E. S. Kissel, A. N. Levy, S. MacCracken, P. R. Mastrandrea, & L. L. White (Eds.), *Climate change 2014: Impacts, adaptation, and vulnerability. Part B: Regional aspects. Contribution of Working Group II to the Fifth Assessment Report of the Intergovernmental Panel on Climate Change* (pp. 1613–1654). Cambridge University Press.
O'Brien, J. A. (2000). *Management information systems; managing information technology in the E-business enterprise* (5th ed.). Mc Graw-Hill Publishing.
OECD. (2013). *The nature of policy change and implementation: A review of different theoretical approaches*. Organisation for Economic Cooperation and Development.
Paschen, P. (2015). *Decent work, green jobs and the sustainable economy: Solution for climate change development*. International Labour Organisation.
Sen, A. 1999. *Development as freedom*. Oxford: Oxford University Press.
Simmons, A. (2021). *Impact of climate change on youth in small island communities Vol 1* (Sustainable development goals series, climate change). Springer Nature.
Slack, N., Chambers, S., Harland, C., Harrison, A., & Johnston, R. (1998). *Operations management* (2nd ed.). Pearson Education Limited, Prentice Hall Publishing.
Thuraya, K. (2015). *Improving the effectiveness of E-learning improvement in the school of engineering at Tripoli University*. Doctoral thesis, University of Huddersfield.
UNDP. (2015). *Human development report; work for human development*. United National Development Programme (UNDP).
UNDP. (2020). Assessment of the socio-economic impact of COVID-19 in Sierra Leone: United Nations Development Programmes; Ellen Johnson-Sirleaf, David Cameron, and Donald Kaberuke 16 June 2020, Freetown.
UNESCO. (2014). *Shaping the future we want: UN Decade of Education for Sustainable Development (2004–2014)*. DESD Monitoring and Evaluation.
UNESCO. (2015). *Not just hot air: Putting climate change education into practice*. Education sector.
Ward, J., & Elvin, R. (1999). *A new framework for managing IT-enabled business change* (pp. 197–221). School of Management.
Williams, D. (2013). *Redesigning Schooling – 3: Principled curriculum design: Redesigning School, the Campaign schoolish vision for education*. SSAT (The School Network) LTD.

5 Curriculum Development as a Pathway for the Implementation of Climate Change Education and Achieving Sustainable Development in Africa

Abstract

This chapter is divided into two sections. The first section discusses the processes and methodologies concern with the review of the curricula and resulting in the development of the CCA for Sustainable Development curricula. The section documents the work of the NCTVA and the implementation process, which led to curricula approval, despite the COVID19 pandemic preventing the Committee from physically reviewing the curricula, training manuals and the assessment procedures together. Its acceptance by NCTVA means that Sierra Leone is the first country in West Africa with approved curricula in Climate Change Education for Sustainable Development.

The second section of the chapter deliberates on the processes and methodologies that led to the development of the Training of Trainers programmes developed for the training instructors and counsellors to deliver the curricula to youth attending training courses run by PIs. The section presents a summary of the methodologies and process involved in training instructors, lecturers, and counsellors to deliver the training courses. The chapter presents an analysis of the views of youth, lecturers and instructors regarding the challenges and gaps confronting the implementation of the training of trainers' programmes.

Keywords

Sustainable Construction · Aquaculture · Climate change · Sustainable development · Youth development · Entrepreneurship · Employment · Climate Change Adaptation Curricula · Climate change adaptation · Training of Trainers in TVET · Sierra Leone · Biodiversity · Technical vocational education training · Conceptual framework

Introduction

Education is viewed as an invaluable medium for enhancing capacity of the youth to address environment and development concerns in Africa (Musau, 2017). It is acknowledged that most of Africa's education and training programmes suffer from low-quality teaching and learning. These programmes also experience inequalities and exclusion at all levels. The governments must invest heavily in child and youth development through appropriate education and health policies and programmes to address the education inequality in the education system. Higher-quality education improves the distribution of skilled workers. Therefore, the state authorities can use this increased supply to build a fairer society in which all people have equal opportunities to contribute to national development.

A. Simmons, *Climate Change Adaptation Framework and Youth Entrepreneurship in West Africa*, Sustainable Development Goals Series, https://doi.org/10.1007/978-3-030-85754-7_5

Presently only the elites benefit from quality education (Musau, 2017).

This chapter is divided into two sections. The first section discusses the processes and methodologies concerned with the review of the curricula culminating in the development of the new Climate Change Education for Sustainable Development curricula which include (a) Sustainable Hospitalities (Tourism and Hotel Management), Sustainable Agriculture (focuses on Aquaculture), Sustainable Construction (Carpentry, Masonry, and Fabrication), Career Guidance and Job Readiness Skills and Entrepreneurship Development Skills. The chapter documents the work of the NCTVA and the implementation process, which led to the approval of the curricula, training manuals and assessment procedures, making Sierra Leone the first country in West Africa to approve a curriculum focusing on climate change education for sustainable development. The second section of the chapter deliberates on the processes and methodologies that produced the Training of Trainers programme developed for instructors and counsellors to deliver the curricula to youth attending the courses. It also includes the trainers' views regarding the challenges in the programme.

Section 1: Methodologies and Processes Used for the CCESD Curriculum Development

Collecting Data from Stakeholders

The consortium utilised multiple research approaches to collect data on the performance of the YEEP for the monitoring and evaluation review process. It conducted desk reviews by synthesising all monthly reports and other documents produced by the consortium and partner agencies. Key informants' interviews were held with the private sector, TVET institutions and other partners to assess partnership processes and relationship, identify what worked well and what did not work well and recommend possible changes. Focus group discussions (FDGs) were conducted with youth to evaluate the quality of the programmes and strategies to enhance the courses' quality and add value to the overall YEEP. These discussions were supported by a series of visits by the consortium's KEs and NKEs to PIs to observe the quality of the content and the methodologies used to deliver the course content.

The multi-search approach provides the research process with the confidence necessary to pursue monitoring and evaluation (M&E) and enabled triangulation. Triangulation increases the validity of claims by determining whether the phenomenon investigated is accurate irrespective of the data collection used (Saunders et al., 2000; Robson, 2002). The consortium's team triangulated the sources of information to build a coherent justification for the data collected (Creswell, 2009). Each data collection method possessed its peculiar strengths and weaknesses (Smith, 1995 in Saunders et al., 2000; Robson, 2002; Angelson et al., 2011) and applied to the specific area of the monitoring and evaluation process which best fits its use:

- Interviews with key informants (head of PIs/instructors) involved in implementing YEEP and developing strategies to deliver it now and in the future.
- FGDs were held among young people from rural and urban areas grouped by each component of the project.
- Case studies with youth on their perspective of the project now and in the future. This method provided the opportunity to extract the lessons learned from the project to make changes and adjustments to policies, strategies and programmes for youth employment through skills training and entrepreneurship in the future.

It is important to note that the methods identified for collecting data relate to the aims and objectives of the monitoring and evaluation process, research questions and the target population (Angelsen et al., 2011). Considerations were made to determine the research methods used in the monitoring evaluation process regarding the relevance of data collected.

Population Sampling for the Monitoring and Evaluation Process

The purpose of a good sampling strategy is to ensure that appropriate data is collected at each level relevant to the nature of the analysis to be implemented by the team conducting the research study (Angelson et al., 2011). This strategy will ensure that data from the different levels can be brought to bear in understanding the social, political and physical environment in which stakeholders operate. In general, the sample is pursued at several different stages of the study (research process) (Robson, 2002; Byrne, 2002 in Angelson et al., 2011). A similar selection process was instituted in the selection of PIs to participate in the data collection process. The names of all the PIs were placed into a hat and randomly selected.

Desk Review

During the M&E process, the team conducted a desk review by collecting, organising and synthesising available information to understand the project's present status. Reports on the performance of the project from inception to date were analysed. The analysis was conducted based on the application of the main OECD/DAC evaluation criteria (relevance, effectiveness, efficiency, sustainability and early indications of success/impact) and examined all the various elements affecting the implementation of the project, including the appropriate organisational arrangements and partnerships (OECD DAC, 2019; Robson, 2002). The consortium conducted a series of consultative discussions among PI representatives during the period under review. Questions were asked on the (a) relevance and design of the project, (b) effectiveness of the project, (c) efficiency of the project implementation process, (d) sustainability of the project, (e) partnership contribution and (f) lessons learned and early indicators of potential successes (OECD DAC, 2019).

Interviews

Unstructured interviews were conducted with NAYCOM senior managers and other government agencies such as the NCTVA to solicit their views and concerns on the sectors and teased out relevant information that was not documented. The interviews provided opportunities for the consortium to re-establish personal contact with officials in government agencies and PIs, thus facilitating trust and cooperation with senior managers. During the COVID-19 pandemic, NKEs were trained on how to administer the tools through telephone communication. Due to the issues relating to the connectivity between Freetown and rural communities in Sierra Leone, a limited number of interviews were done online (Saunders et al., 2000; Robson, 2002). Interviews with NAYCOM, NCTVA and other senior officials from government, private sector and other public sector institutions were conducted by the project leaders (check Appendix 11 for sample of interview questions).

Questionnaires

The questionnaire was viewed as the preferred data collection tool to collect data from head of PIs and instructors/counsellors and youth/beneficiaries of the three components of the YEEP during the COVID-19 pandemic. To reduce the incidence of contamination of data, the NKEs and KEs advised PIs to administer the tools through personal emails and requested participants to return the emails to them as soon as they completed filling in the forms (Robson, 2002). However, due to the slow responses from PIs, the consortium staff administered the questionnaires among youth and instructors and returned the FGDs and questionnaires data forms to the office (check Appendix 1V for sample of questionnaire).

Observations

A series of observations were carried out by the project lead, KEs and NKEs at PIs to examine the performance of heads of PIs and instructors/counsellors within their respective institutions, including the delivery of actual training courses, the interaction between youth and instructors and the effectiveness of the methodologies utilised to deliver the training (Robson, 2002). The KEs and NKE ensured that PI instructors and youth were aware of their presence and the nature of the work. The disadvantage of informing them before a visit is a failure to participate or altering of behaviour in such a way as to present invalid information, which the consortium has no control over. Observation is still a meaningful way to gain insights, even if it is just an awareness of behaviour modification, once used with other data collection tools.

Case Studies

Case studies are bounded by time and activity, and researchers collect detailed information using a multitude of data collection procedures over a sustained period (Stake, 1995 in Creswell, 2009). Qualitative case study methodology provides tools for the consortium to study complex phenomena within their contexts (Baxter & Jack, 2008). Stake (1995 in Creswell, 2009) and Yin's (2003) approach to case study development based on a constructivist paradigm, i.e. truth is relative and dependent on one's perspective, was used. A set of criteria were developed in collaboration with PI officials to guide the selection of two learners to participate in the project. Sample questions were also developed and sent to PI heads to administer the questions among learners who were selected for case studies. The consortium prepared the case studies based on the data submitted by the partner institutions. (Check Appendix III for sample of questions for the case studies).

Utilising Thematic Analysis Tools to Analyse FGDs and Interview Data

Thematic analysis is a phenomenological method for identifying and analysing patterns in qualitative data (Guest et al., 2012; Clarke & Braun, 2013). Some researchers identify the approach as a qualitative analytical method rather than a full-fledged methodology (Clarke & Braun, 2013). The thematic analysis focuses on what is said rather than how it was said, allowing key themes to emerge from the data. It works with a range of research questions, can be used to analyse different data types, works with a range of datasets and can be applied to produce data-driven or theory-driven analysis. While there is no accepted, standardised approach for carrying out thematic analysis, in Clarke and Braun (2013) put forward a systematic approach to implementing thematic analysis. This process comprised of a six-step analytic approach, including (1) familiarisation with the data, (2) generating initial codes, (3) searching for themes, (4) reviewing themes, (5) defining and naming themes and (6) producing the report (Simmons, 2021).

Validation Workshops

A series of participatory workshops were organised involving all stakeholders to validate the participants' responses by triangulation, providing opportunities for the consortium to cross-check the data collected from the various data collection exercises. It also provided opportunities for the multiple stakeholders to interrogate and clarify the data collected from the various data collection instruments. These workshops were implemented at critical stages throughout the duration of the project.

In summary, the thematic analysis brought together data from the interviews, questionnaires and focus group discussion datasets relevant to a theme. This enabled those writing up the reports to ensure a clear, concise and comprehensible depiction of the experience of those who contributed. The thematic analysis transformed the individual life stories of the participants into a body of accessible information concerning a life experience. The data presented central themes with sub-themes on the perceptions and experiences of NAYCOM senior managers, PI heads and instructors and youth relating to the development and implementation of the YEEP (Clarke & Braun, 2013).

Reviewing the TVET Curricula and Training Manuals in Sierra Leone

The methodologies above were used to ensure the consortium's YEEP implementation process was highly participatory for all stakeholders and their institutions involved in reviewing the curricula of TVET institutions (Simmons, 2021;

William, 2013). During the curriculum review process, desk reviews and several interviews were conducted to ensure that the gap between the skills provided by training institutions and those required by the job market is compatible (UNDP, 2015; Carter, 2008). The review was done to aid in forecasting labour market needs and trends, providing up-to-date information on skills needs and occupational areas presenting hiring challenges. Furthermore, it also prevented any mismatch of skills training and aligned education provider programmes according to the current and future needs of the labour market.

The review of the curricula was implemented in collaboration with the NCTVA, the Ministry of Higher Education, the business sector and other stakeholder institutions. The consortium worked on developing the new curricula and TVET training manuals and tools to enhance the quality of the training and match the competencies required by employers. In line with the priority areas in the Government of Sierra Leone's medium-term National Development Plan (MTNDP, 2019–2023), several goals were identified for its flagship programme to promote free quality school education and to provide a solid base to enhance human capital development and facilitate the transformation of the economy (Government of Sierra Leone, 2019; IMF, 2019). Three of the four National Development Goals in the MTNDP are linked to YEEP, which is (1) a diversified, resilient green economy;(2) a nation with educated, empowered, and healthy citizens capable of realising their fullest potential; and (3) a competitive economy with a well-developed infrastructure. YEEP supports the MTNDP goals as they enable young people to drive inclusive green growth that is sustainable.

Climate Change Education for Sustainable Development Programme

The curricula development review process was implemented in line with the UNESCO model for the development of Climate Change Education for Sustainable Development (CCESD) programme curricula (UNESCO, 2010) and the Climate Change Adaptation Framework, which was developed for SIDS (Simmons, 2021). This programme uses innovative educational approaches to help a broader audience (focusing on youth) to understand, address, mitigate and adapt to the impacts of climate change. The CCESD initiative encourages changes in the attitudes and behaviours of youth required to put the world on a more sustainable development pathway (UNESCO, 2014). This framework contributes towards building a new generation of climate change-knowledgeable citizens who are motivated to take actions to protect their environment.

UNESCO (2014) reaffirmed its commitment to promote and improve inventive teaching methods to incorporate quality CCSED in the education processes through interdisciplinary practices, science education, whole-school approaches, TVET and education on disaster risk reduction. Integrating CCESD into curricula without compromising already overstretched programmes of study is a challenge (UNESCO, 2010). The main action to employ when developing CCESD in the curricula is through a transdisciplinary manner where CCESD is infused in existing subject areas such as science, citizenship education, geography, human rights education and languages, rather than implemented as a separate entity added to the already congested curriculum (UNESCO, 2014). On this basis, a set of five skills courses were selected to be reviewed, including Carpentry, Hospitality and Tourism and Fish Farming. The consortium agreed in collaboration with NAYCOM to explore the development of new courses in Career Guidance and Job Readiness Skills and Entrepreneurship to enhance the competencies of youth in Sierra Leone.

The skills selected linked to the Government of Sierra Leone's MTNDP 2019–2023 (IMF, 2019) and were supported by a consultative process implemented by the consortium during YEEP's inception stage. The three areas selected for the Skills for Employment component were considered high-priority employment-generating drivers to propel the economic development of Sierra Leone. The consortium and NAYCOM organised a further comprehensive review of the course content, qualification of instructors and quality of training facilities. The courses offered in the selected skills were reassessed against the type of skills set and competencies required by employers in Sierra Leone to create sustainable livelihoods in the future. New courses were proposed to align training according to the needs of the labour market. The training methods and equipment were assessed during the visits to the selected TVET institutions. During these visits, it was noted that there was an absence of modern equipment, tools and training materials to train the youths in these participating TVET institutions. The tools and equipment utilised for teaching youth in these institutions were either incredibly old, not operational or non-existent. An updated list of equipment, tools and training material was compiled by the consortium and submitted to NAYCOM for consideration. Through AfDB, YEEP provided $150,000 for the National Youth Commission to purchase duty-free tools and equipment to support the implementation of the TVET training programme in the respective participating institutions.

The Hotel and Tourism Management Course

During a skills audit consultation, the lack of foreign language skills ranked the highest on the skills gap for youth working in the Hotel and Tourism Management sector in Sierra Leone. French was the only foreign language in the curriculum. According to a National Tourist Board representative, the language barrier has restricted Sierra Leone tourism to most countries in Europe and the Arab regions. Based on the recommendation from National Tourist Board, the consortium recommended that basic German, Italian and Arabic should be added as elective courses for students pursuing studies leading to certificates or diplomas. The stakeholders recommended casino gaming and introduction to spas and physical fitness as new skills to boost the sector. The present certificate course only offers training in rudimentary ICT. A new module in the property management system (PMS) was recommended to train the youth already participating in training for this sector. As a front-office tool, the PMS is at the centre of hotel operations, helping to boost the efficiency of the tourism facility by simplifying the check-in and check-out process. Opera and Maestro software packages were also recommended for youth to be trained in to increase their efficiency and effectiveness on the job.

Current Training Methods and Equipment

The lack of equipment, tools and materials to conduct training at the TVET institutions severely impedes the quality of the graduates produced (UNDP, 2015). Even if qualified teachers are provided in the institutions, graduates will not function effectively in the workplace without the required teaching and learning tools and equipment. They would require additional training by industry 'to get them up to speed' costing businesses unnecessary time and money. Hoteliers and other managers working in the sector complained that the graduates produced by these agencies were of low quality and cannot function beyond basic operational levels. Some graduates could use rudimentary equipment such as vacuum cleaners, microwave ovens, washers and dryers; lack of exposure to equipment required by the workplace limits their ability to work effectively.

The students pursuing Hotel Management and Tourism Programme at the School for Tourism and Hotel Management, MMCET Brookfield, complained of only gaining opportunities to use modern equipment and tools during professional attachments or internships. In some instances, this has resulted in the extension of their internship experience from 1 to 3 months; interns are only limited to a specific type of training rather than working across different departments within hotel management. Only four computers are on campus to train 200 students in basic ICT.

The Tourism Course

The course content mainly focused on graduates becoming certified tour guides which is a major shift from the tourism management training course. The curriculum provides the participant with the knowledge of the types of tourism in Sierra Leone, tourist attraction centres and their locations in Sierra Leone, basic knowledge of the operations of travel agencies and tour businesses. Certified tour guides can be employed by both the private sector and government institutions engaged in the industry as well as establishing their own businesses. A 3-month foreign language skills training course is included in the programme to satisfy the recommendation by the National Tourist Board. It is vital for improving communication and increasing opportunities when taking guests on tours and excursions across the country.

As the tourism department of the MMCET lacked equipment, tools and learning materials to prepare students for employment in the tourism sector, practical excursions and materials on historical sites, cultural occasions, beaches and islands were recommended as part of the content for the training to bolster the quality of graduates. The training courses incorporated community tours and city tours to areas of tourist attractions. Tools and learning materials (booklets on historical sites, etc.) were produced by students as part of the training exercises to support future training of youth, which minimised the costs required by institutions to produce new resources. The National Tourist Guide Association gave technical assistance, which increased the course's credibility and the capabilities of graduating tour guides.

To complement the YEEP's national climate change education curriculum, the existing Tourism and Hotel Management courses were combined and modified to Sustainable Hospitalities by including a module in Climate Change and Sustainable Development. The additional modules for the Sustainable Hospitality course include types of tourism in Sierra Leone, managing tourism-related properties, organising and managing tour operations and excursions, becoming a certified tour guide, knowledge of tourist attraction centres and their locations in Sierra Leone, e-travel and a basic understanding of the operations of travel agency and tour businesses. These modules were in addition to a 3-month foreign language skills course vital while taking tourists on tours and excursions. MMCET launched a consultative stakeholder's workshop to further unpacked the new curriculum and reorganised the various departments of the hotel school around the modules in the new curriculum.

Carpentry Training in TVET Institutions

Most TVET institutions offer training in Carpentry and Joinery in Sierra Leone. The theory and application of carpentry skills development include an in-depth study of materials used in the carpentry field, foundations and floor layout and framing of buildings, ceiling framing, roof framing, installation of windows and doors and stairs layout. This training is no longer relevant because there have been colossal changes in the way buildings are constructed in Sierra Leone.

Climate change impacts housing and communities across West Africa, resulting in more severe storms, flooding and landslides and their devastating effects in communities. Stakeholders of the construction industry recommended the inclusion of a Sierra Leonean Timber Handbook, categorising the different species of timber found in Sierra Leone, their local and botanical/trade names and classification and building codes. The handbook should also distinguish between types of wood used for furniture and construction products, those produced naturally and those from agroforestry. Recommendations from the private sector also highlighted the need to include aluminium fabrication as part of the construction module. Aluminium has become widely used in West Africa. Currently trained masons and carpenters find it challenging to gain employment as they lack the skills to produce and install aluminium windows, doors and other products. The young people completing the certificate training course were therefore trained on building and installing these products.

Current Training Methods and Equipment

On the initial review of the PIs, the equipment and training facilities at GTI were found to be outdated. There were inadequate teaching and learning materials, including the absence of e-books, modern textbooks and other forms of digital learning aids to assist youth in capturing the essence of the learning process, utilising modern and appropriate tools. The methodologies used by the institutions to train the youth in the TVET skills were not practical-oriented, user-friendly or focused, with limited real-world practical case studies and activities inside and outside of the classroom. The consortium put together a list of modern tools and equipment required to train youth to acquire construction skills and recommended using ICT and audio-visual teaching methods in and out of the classroom to sustain interest by learners using more colourful, animated presentations and real-life examples in the learning environment.

Traditionally, construction education focused primarily on the management, materials, methods and technical aspects of the construction industry, while environment and sustainable development issues were ignored. However, as the world becomes more aware of the impact of climate change on the environment's degradation, we understand that construction companies and personnel significantly affect the environment. Therefore, there is a need for a more holistic approach to constructing communities, including strategies to be implemented to deal with waste management, energy-efficient building design and the overall physical environment (sustainable practices and value systems).

Sustainable Construction Course

Sustainable construction is viewed as a system using expertise, materials and methods to construct and/or maintain a building structure that meets the needs of the present without compromising the ability of future generations to meet their own needs (UN, 1987). The Sustainable Construction course educates youth on developing green building attributes and benefits and provides them with strategies to improve the environmental challenges that affect the built environment. It also focuses on enhancing the efficiencies and sustainable use of energy in communities by acquiring sustainable development practices. Through lessons, audits and activities, students cover topics including water and energy efficiency and environmental quality related to building construction, operations and maintenance.

Alongside the timber handbook and aluminium fabrication, training in masonry and creating sustainable block building for construction (recycling plastics) was also included in the new Sustainable Construction course. Climate change adaptation is taught as part of the certificate and diploma training courses, with ICT modules complementing the training.

Aquaculture in Sierra Leone

Inland fish farming in Sierra Leone started in 1976 with the establishment of a government fish breeding station at Makali in Tonkolili District. This initiative was pursued by the government's Rural Development Project Aquaculture experimental station in Bo. The fish produced from these fishponds were mainly consumed in rural and non-coastal communities. The Government of Sierra Leone expressed its desire to develop aquaculture as part of its food security strategy by providing affordable fish protein to the poor urban and rural population (IMF, 2019). Local freshwater fish already provide a rich source of animal protein in rural settlements located along rivers and in wetland area in Sierra Leone.

Fisheries Management in Sierra Leone

The fishery sector in Sierra Leone is impacted negatively by climate change. Fishery is one of the economic sectors that urgently requires climate change adaptation (CCA) support to reduce the impact of climate change on the sector. Alternative and complementary livelihood activities are required to build resilience of society to respond sustainably. Ministry of Fisheries and Marine Resources (MFMR) has responsibility for all fisheries in Sierra Leone. The fishing industry in Sierra Leone consists of two sectors, the industrial and the artisanal, the latter comprising marine artisanal, inland artisanal and aquaculture. Fish is an important component of the diet, which supplies approximately 80% of animal protein consumption in Sierra Leone. Aquaculture is concentrated in provinces in the south (mainly in Bo and, to a lesser extent, Moyamba and Pujehun), north (mainly in Tonkolili) and east (in Kailahun, Kenema and Kono). A culture of fish in earthen ponds is the most common system of aquaculture. The principal cultured species is the *Oreochromis niloticus*, but the catfish *Clarias gariepinus* and *Heterobranchus* spp. are also cultured.

Although no current data exists on the number of persons involved in the fisheries sector in Sierra Leone, it is estimated that over 40% of the fisherfolk are youth. Many youths are also employed in fisheries-related businesses such as boat building, servicing engines for fishing boats and shipping fishes locally and abroad. Therefore, any reduction in the fish stock will further impact the livelihoods of the youth population. Changes in migration patterns and depth of fish stocks due to climate change are two of the main factors affecting the distribution and availability of fishes. The fish sector is hugely exploited by foreign trawlers responsible for scraping the coast water of its fishes. The destruction of the mangrove swamps robs the sea of plankton and young fish stock, limiting food supplies for more mature fishes. Research shows that loss of coral reef habitat is having damaging effects on the coastal fisheries sector (Pratchett et al., 2009), where reef-based livelihoods and tourism activities are important to the economic development and well-being of persons living in coastal communities around Freetown and the peninsula (Hernandez-Delgado, 2015).

Training Methods and Equipment

Although some training was provided in Aquaculture by TVET institutions in Sierra Leone, there was no evidence of adequate training tools and equipment to train the youth in this new occupation. The PI, Njala University, lead in aquaculture, providing training to youth from certificate to masters' degree levels. The institution possesses the expertise to develop the curricula and training in this skills area even during the initial review. However, the course content provided was standard, and an approved curriculum for aquaculture was absent in Sierra Leone. The consortium worked with the training institutions and the private sector to list the required equipment, materials and tools and the curriculum for submission to NAYCOM and the NCTVA for feedback and approval.

Sustainable Aquaculture Course

The course developed for the YEEP includes the following topics: the development of fisheries in Sierra Leone; the relationship between hydrography and fisheries; the essential requirements for establishing fish farms; the construction procedures of a pond; the culture of brackish and freshwater fish; the harvesting process of fish in ponds; and fish preservation, processing and distribution methods. The new curriculum also included additional modules such as climate change, the environment and sustainable development, entrepreneurship development and the linkages to the overall development of the economy. The Aquaculture training course was implemented as an independent certificate course or part of the Sustainable Agriculture or Climate-Smart Agriculture diploma or degree course.

Agriculture in Sierra Leone

Of all the economic sectors, agriculture is still the largest employer of people globally, with a growing workforce of over one billion people, i.e. one in every three workers (Poschen, 2015). It is also the sector with the largest concentration of poor people, of which two-thirds live in rural areas. Research shows that agriculture is the largest emitter of greenhouse gases and is the most vulnerable to climate change globally (IPCC, 2014a, b). It is also one of the most prominent users and polluters of water and a cause of land biodiversity degradation in most developing countries (Poschen, 2015; Sustainable Development Solutions Network – Youth, 2020; UNEP, 2021).

Climate change impacts agriculture through increased variability, including temperature rise and rainfall intensity, resulting in the flooding of low-lying areas and landslides. Such frequency and intensity of extreme weather events such as hurricanes and storms caused changes in rain patterns and water availability, destruction of ecosystems and decreasing productivity in certain areas due to reduced soil fertility and soil loss. Yet, despite all the challenges associated with agriculture production due to the severity of climate change, food production needs to be increased by up to 70% to feed the world's growing population adequately. Deliberate and concerted programme training of farmers, specifically small-scale farmers, in developing countries such as Sierra Leone, in productive farming methods, is pivotal to promote sustainable agriculture (Government of Sierra leone, 2019). The advantage of the low-impact method of intensive training for farmers is that it is labour intensive and would increase the available jobs for young people in the short to medium term. The Government of Sierra Leone, through the NAYCOM, has already secured large parcels of land where it is implementing viable 'youth in agriculture' projects across the country. However, it lacks a sustainable agriculture curriculum supported by a comprehensive training programme to enhance small farmers' skills, knowledge and capacities. The training programme should be supported by investment and extension services to strengthen rural infrastructure and develop the non-farm rural economy (Poschen, 2015; Sustainable Development Solutions Network – Youth, 2020; UNDP, 2020; UNEP, 2021).

Climate-Smart/Sustainable Agriculture Course

The training consists of classroom training run by TVETs, supported by practical training intervals in rural communities. The practical aspects of the course would be conducted on NAYCOM's farms and private farms, where available, to provide insight to youth on how these farms are managed and operationalised. Students would be placed on a 3-month professional attachment on shortlisted private companies' farms offering hands-on practical experience. The sustainable agriculture training modules include cropping, soil fertility, management and preparation, improving farming methods, marketing of products, entrepreneurship and links to the other economic sectors, the use and management of pesticides and other hazardous chemicals, food security and environmental impact of agriculture.

The TVET institutions responsible for implementing the training course would have to ensure that the youth placed on the attachments are visited monthly during their placement. Completing practical work experience is a prerequisite in the final determination of successful completion of the accredited certificate and/or diploma course in Sustainable Agriculture. Supervisors from the partner agency would fill in a logbook and submit M&E reports for each youth during each field visit. This is to foster continual assessment, maintain communication, provide support, identify any gaps in the student's progress and immediately deal with any challenges that occur during the internship.

Curricula Assessment Model

One of the objectives of education is to contribute to society's economic growth and sustainability. To be effective and responsive, education must be high quality and demanded by the economy and society (Simmons, 2021; William, 2013; Olaniyan & Okemakinde, 2008). There is a need for policymakers, educators and representatives of training institutions to link education programmes and the skills provided with market needs. This is because better-quality education will help people develop a clearer understanding of coping with and responding to changing climatic conditions and other development calamities (ILO, 2010; Paschen, 2015; UNDP, 2015). Therefore, implementing a rigorous assessment mechanism to assess the performance and relevance of the curricula will provide indications on the effectiveness, efficacy and efficiency of the equality of education provided to youth.

Assessment is integral to the teaching and learning process; it provides information that can help improve youth learning and support educators in delivering learning content. Over the past decade, the assessment of vocational education and training has shifted from curriculum content to outcomes or competencies. Competency-based assessment (CBA) has emerged as a viable assessment system for TVET education, i.e. utilising appropriate methods to gather evidence of competence and interpret the evidence against the competency standards (Mukhtara & Ahmadb, 2014, 2015).

Operationalisation of the YEEP TVET Assessment Framework

Considerable effort is needed to ensure that the skills development paradigm delivers both the quality and quantity of the training required. This involves the 'adequate supply of quality teachers, training instructors, trainers, director of training institutions and master craft-persons to take an apprentice; the provision of opportunities for them periodically upgrade their own skills and conditions of work comparable with those in in the industry so to attract the most talented staff' (ILO, 2010 pg27).

Assessment for learning (AFL) is an approach to teaching and learning that generates reaction, which is then used to enhance students' performance. Students become more engaged in the learning process, and from this increase confidence and insights in what they are supposed to learn and at the standards required. AFL is important in education and is an essential mechanism of the CBA process in TVETs. Mukhtara and Ahmad (2014, 2015) state that implementing CBA in vocational programmes requires teachers to practice AFL through the education process. A comprehensive assessment process was implemented by the consortium involving all YEEP stakeholders in the assessment process:

- **Formative assessments** take place in the training environment to provide information on a trainees' progress. This data was used to reinforce and complement individual learning. Formative assessment is an essential component of the training procedure, and the results are used to set training goals and provide practical responses to the learner and the learning process.
- **Summative assessment** is used for awarding formal qualifications when a student is evaluated against a specification or standard. It may take place at different points of the training process, but the student must have been informed that his/her assessment results will be logged, and the necessary quality must be ensured.
- **Diagnostic assessment** is used to determine trainee's strengths and weaknesses, find or tailor a training programme or assign a specific group.

Assessment Procedures

The YEEP assessment model recommended by the consortium was linked to the assessment procedures and process used by the NCTVA in Sierra Leone to run throughout the implementation phase. NCTVA assessment process enforces national standards and processes. It is trustworthy and reflective of the nationally recognised standards of competence defined by the industry and professionals within specified occupational fields. The assessment model goes through the following procedures: observation, product evaluation and questioning.

Observation: The trainer will observe the student as they carry out tasks defined in the standards for the qualification in the workplace or any other place where the candidate is undertaking practical activities. The trainer will need to plan observation in advance to take advantage of any skills or activities that occur naturally in the learning environment and make the best use of the available resources. A checklist of sub-competencies that the trainee should perform during the exercise under observation will be used. Trainees receive adequate information before the date set for the visit.

Product Evaluation: Trainees individually or in groups work towards attaining their qualification by creating a product up to a chosen stage in the product creation process, such as the construction phase. The range of things that can count as a 'product' is extensive, including (a) design assignment and/or (b) a reflective report supported by portfolio evidence from workplace experience. The critical consideration here is that the examiner assesses the evaluation of a completed product.

Questioning: Questioning assesses trainees' knowledge and comprehension and how they apply their knowledge through reasoning, planning, analysing and evaluating. Questions may be oral or written. Oral questions are generally used to (a) confirm knowledge and understanding where it is not apparent from performance, (b) address gaps in knowledge and understanding in performance-based units and (c) validate evidence by asking learners to explain part of the evidence or describe the process of producing it. Questioning assessment includes (a) written tests/assignments, (b) oral questioning or tests, (c) professional discussions, (d) performance assessment, (e) portfolios, (f) projects, (g) aural assessment, (h) case studies, (i) roleplays, (j) simulations, (k) self-reports, (l) interviews, (m) questionnaires and (n) third-party/witness testimony. Trainers and instructors were trained during the Training of Trainers to use a wide range of methodologies to deliver the course's content and assess the competencies of youth and adult learners throughout the YEEP implementation process.

Practical Supervision and Evaluation Procedure

Practical supervision is highlighted as an important part of the certification process of TVET programme in Sierra Leone. For example, the MMCET tourism assessment process states that before a candidate is qualified to sit the final examination, the candidate should have (a) attended a minimum of 75% in class (all sessions), 85% in practical activities, with practical skills fully documented, (b) obtained no less than 50% in the examination conducted in all semesters and (c) took part in excursions, practical duties and project field trips as deemed appropriate. In terms of the final grade, the examination accounts for 70%, and the continuous assessments (practical) account for 30%. Due to the focus driven towards CBA, private sector agencies recommend 60:40 for examination and practical skills, respectively.

Supervision of learner on practical attachment and internship is an important part of the assessment process of TVET programmes. The supervisors from TVET institutions would ensure that learners placed on attachment and internship are visited monthly to observe them on their placement. Completing practical work experience is a prerequisite in determining whether the learner successfully completed the programme. The internship placement allows the supervisor from the training institution to liaise with the private/public sector agencies on the skills required in the workplace. Over 40% of graduates receive employment with companies because of their performance on the internship placement programmes. Several youths, specifically those in the Sustainable Hospitalities and Sustainable Construction Certificate courses, were disappointed when they could not participate in internship placements due to the COVID-19 pandemic.

Reflection on the Curricula

The Government of Sierra Leone is putting measures in place to leapfrog into the knowledge economy. Developing the new set of competency-based climate change education for sustainable development curricula is its first step towards ensuring the achievement of its sustainable development goals. On approval of the new curricula by the NCTVA, Sierra Leone became the first country in West Africa with a set of proved curricula designed on the UNESCO Climate Change Education for Sustainable Development (CCESD) framework. The CCESD prepares people to plan for, cope with and find solutions for issues that threaten their sustainability (UNESCO, 2014). The CCESD empowers people to make informed decisions for economic viability and just society for present and future generation while respecting cultural diversity. Young people recognise the importance of CCESD as a vital development model; they view it as education for future sustainable living, i.e. as a mechanism that would enable them to develop appropriate skills and knowledge to build resilience to reduce the impact of climate change and other development calamities on their communities and to create sustainable livelihoods (UNESCO, 2014; Simmons, 2021).

Technology is an important driver for enhancing and advancing development at national and regional levels across the globe (Earl, 2001; Hickman, 2002; Poschen, 2015; UNDP, 2015). Like most LDCs in Africa, Sierra Leone lacks the appropriate Internet connectivity to drive equitable development across the country. The slow pace of technological innovations in Sierra Leone and the West African region contributes to the retardation of the national sustainable development goals.

The Government of Sierra Leone is making concerted efforts to create an enabling environment by offering tax relief and other incentives to entice global private sector companies to invest in the country's development. Such investment, if managed efficiently, would result in rapid growth in the economy. However, the demand for higher-skilled workers by these companies will not be satisfied currently by Sierra Leoneans due to the gap and mismatch in skills and competencies of the workforce (Carter, 2008; UNDP, 2015).

Most of the TVET institutions in Sierra Leone do not have modern nor reliable Internet facilities, tools and equipment to maintain functional strategic and operational IT systems to drive their training programmes. In most rural communities, Internet connectivity is either nil or ineffective, making it extremely challenging for universities, colleges and TVET centres to deliver e-learning programmes to learners pursuing YEEP training courses. Most training institutions also lack modern computers to train the youth. An assessment of students' ICT capabilities showed that over 80% do not have computers or facilities to access e-learning. Although there were efforts on the part of the NAYCOM to provide some basic modern tools and equipment to the training centres, these tools and equipment were severely limited compared to the number of trainees signing up for these courses. Through TVET, the Government of Sierra Leone is developing inclusive practice and policies for the poor with disabilities and their families, marginalised and unemployed youth, disadvantaged women, ex-child soldiers and other marginalised and disadvantaged groups to access the training (Milio et al., 2014). However, due to financial, geographical and structural challenges, most young people are still experiencing difficulties accessing the programme.

Besides providing opportunities for a robust interface between the world of work and the world of education and training, the project offered a new set of curricula, training methods, content, tools and equipment to TVET institutions. These assisted in building the resilience of youth and the institutional capacity of TVET institutions to adapt to the impact of climate change and other development calamities. The model adopts the hybrid development approach linking top-down governmental institutions with bottom-up NGOs and other PIs (OECD, 2013; Stochowiak et al., 2016). The hybrid approach boosts the capacity of NCTVA to work with a range of development institutions and professional groups to enhance the management and governance of TVET institutions to deliver quality education at national and local levels (OECD, 2013; Stochowiak et al., 2016).

NCTVA lacks the human and financial capital to fulfil its goal and mandate of strengthening the capacity of TVET institutions and provide curricula relevant to the country's development needs. Using the consortium's limited resources, NCTVA established a viable and functional advisory body involving representatives from the private and public sectors, TVET institutions and technical experts relating to the various occupational areas to provide technical advice and feedback to the agency on the new curricula. NCTVA worked with the consortium and its advisory group through a participatory consultative process leading up to the approval of the Climate Change Education for Sustainable Development curricula (https://nctva.org/).

The development and implementation of a sustainable education process are essential to achieving high-quality education in the future (ILO, 2010; Poschen, 2015). It is no longer enough to train youth to meet current needs; they need to develop skills, knowledge and competencies that support entrepreneurship, resilience, innovation in enterprise development and their transition to sustainable livelihoods. Therefore, the assessment process and the data collection tools discussed above were supported by the management of NAYCOM because they provide a method for the PIs to assess the quality of the institution's training. Through NCTVA, NAYCOM, the Ministry of Higher Education and training providers, the government must continue to provide resources to ensure youth continue to have access to quality TVET training programmes moving forward.

The COVID-19 pandemic created severe disruptions to the project, specifically the NCTVA curricula review process, as from April to June 2020, education establishments were closed during the national lockdown. NCTVA utilised the virtual communication platform Zoom to create conference calls for stakeholders to review the curricula to navigate the initial difficulties of setting up an in-person meeting. However, the poor Internet connectivity of some rural areas in Sierra Leone prevented some technical experts from participating in these consultative meetings. NCTVA was very proactive in sending hard copies of the documents through local couriers to all experts to assess and submit their feedback. This ensured those who could not attend the meeting still could share their expertise and validate the suggestions, which led to the approval of the curricula.

Additional curricula and training manuals in entrepreneurship development and Career Guidance and Job Readiness Skills (CGJRS) were developed and submitted to NCTVA for approval. There was no curriculum on CGJRS as a special area of study; the management of TVET institutions requested that NAYCOM and the consortium work with them to develop a standalone curriculum and training manual and implement a Training of Trainers programme to train counsellors to deliver the course. The private sector expressed similar views that persons completing their tertiary level training at TVET institutions had very poor soft skills such as time management and writing business letters. These views were incorporated in the approved programme to ensure competent and proactive professionals completing CGJRS courses and associated certificate, diplomas and degrees.

The COVID-19 pandemic severely disrupted the internship and apprenticeship placements of youth, resulting in placements of approximately one-third of the graduates. TVET institutions were very proactive and extended their training programmes by 3 months to provide additional practical training in the use of tools and equipment to young people to reduce gaps in their training, provide additional knowledge and skills and boost the overall quality of their practical training experiences (UNDP, 2015; Poschen, 2015; ILO, 2010; Carter, 2008).

Section 2: Analysis of the Training of Trainers Scheme for Instructors and Counsellors

The Training of Trainers (TOTs) programme for instructors and counsellors of TVET institutions was part of the strategy utilised by the consortium to enhance the capacity of PIs. It aimed to conduct effective and result-oriented training of TVET instructors and counsellors who would train youth in skills for employment, entrepreneurship and CGJRS in Sierra Leone. The content and the methodologies of the TOTs were influenced by the data collected from the capacity and institutional assessment and stakeholders' consultative processes conducted during the interim stages of the project. The assessment outcomes outlined gaps; skills mismatch; and limited capabilities of the TVET institutions, particularly the training methodologies, the lack of modern industrial tools and equipment and the absence of viable IT systems to support decision-making (ILO, 2010; UNDP, 2015; Poschen, 2015).

The success of the YEEP project was hugely dependent on the efforts of PIs to drive the project forward; training PIs in modern course content, appropriate teaching methodologies and providing them with the required tools and equipment were prerequisites to implement YEEP successfully. Researchers recognise the training of TVET institutions' training as essential for reforming the TVET educational system to support the implementation of CCESD initiatives (UNESCO, 2014; Puschen, 2015). Interviews with various stakeholders endorsed the view that professional development initiatives can improve the quality of teaching and learning sustainably, increasing the effectiveness of education and training and adding value for learners, teachers and trainers (UNESCO, 2014).

In essence, TOTs enable TVET teachers and trainers to deliver quality training receptive to emerging labour skills.

TOTs for the skills for employment programmes (component 1) were conducted at three institutions implementing the YEEP. At GTI, 13 instructors were selected and participated in the TOTs for the Sustainable Construction certificate course. Participants were trained in (a) sustainable development with emphasis on the impact of climate change on the economy and other sectors, (b) new teaching methods and techniques, (c) operation of updated tools and equipment, d practical application of Aluminium Fabrication and (e) laying out of a conducive training building structure. The TVET instructors congratulated the consortium for providing such viable knowledge and skills regarding TVET training. The instructors claimed that they have been using the old teaching method, which is tedious and boring. They now realised there are many new methodologies and new knowledge they could utilise with young people, including audio-visual support.

Instructors at MMCET participated in a 4-day TOT workshop for the Sustainable Hospitality certificate course. After reviewing the Tourism and Hotel Management curriculum, the Tourism Department at MMCT expressed the need to reassess the internship and industrial attachment timing following the tourist calendar submitted by the Ministry of Tourism. Initially, students were sent on internships during the tourist off-season, which was not beneficial for their training. At the end of the training workshop, the MMCET management decided to establish an e-Travel Lab to train youth to use the Amadeus and basic computing skills to enhance efficiency and decision-making in the travel and tours sector of the tourism industry (Earl, 2001). The instructors from the MMCET School of Tourism food department stated that although they train students to work in hotels and restaurants as chefs, they were not aware of the requirements for producing competent chefs to work in the industry. 'We are lecturers in the Food Department, but we are not knowledgeable on the skills to prepare the most recent dishes used in hotels, so we appreciate the training and demonstrations by the expert', Gloria Green, a food technology instructor, underscored during the M&E session. This statement is an excellent example of the mismatch in the skills taught in training and education institutions and the skills required in the workplace (Carter, 2008; ILO, 2010).

The third and final TOT for component 1 was held at Njala University, Mokonde, where 18 alumni and lecturers from the Aquaculture Department, Faculty of Natural Resources, participated in the 5-day TOT. Besides providing training to instructors in sustainable development and the impact of climate change, training methodologies and strategies for constructing fishpond and sustainable fish processing techniques, instructors were also trained in entrepreneurship in aquaculture. During the TOT, the participants launched the Association for Aquaculture and Fisheries in Sierra Leone, which the university hosts; national fisheries practitioners, businesses, and institutions are registering with the organisation. A concept note was developed to secure funds to support the institution's aquaculture projects. This concept note has since evolved into a project proposal, which received funding from donor agencies to support the association's activities. 'I have been a blind man all this while, but after having gone through the entrepreneurship session, now I can see clearly the opportunities that lie ahead of me', Mr. C, a trainer at the TOT on Aquaculture, said.

TOT Methodologies

The trainers were trained on a range of methodologies to deliver training to youth, reducing monotony and increasing product knowledge and skills. The training focused on promoting participatory learning, ensuring that the training process was exciting and productive using various methods. Instructors were also trained on using technologies such as the Internet to enhance the training content, knowledge and skills.

The following training methods were employed during the ToT:

- **Brainstorming**: This activity generated initial interest by involving all the trainers in the training session. The session also involved sharing ideas and approaches through small groups and whole-class actions.
- **Interactive Talks**: This strategy encouraged trainers to be active and analytical in their learning approach. It allowed trainers to be motivated and curious about new knowledge and skills by asking questions and exploring alternative strategies to resolve issues and challenges.
- **Group Discussion**: This method enabled effective participatory learning among small groups of trainers and provided opportunities for them to present work to their training peers. It also enhanced the confidence of trainers who were more comfortable speaking in smaller group settings.
- **Panel Discussion**: This method involved resource persons from the private sector agencies and the Government Ministries and Departments in the training process. These experts conveyed knowledge to instructors on their specific areas of competencies, and the discussions were tailored to enhance interactions between trainers and agencies to assist in promoting participatory learning.
- **Classroom Practical**: This methodology reinforced the learning experience through classroom practice. It supplemented the knowledge input given to trainers through the lecture method. The trainers were excited to participate in the practical classroom exercises. Originally, the project was to train 2700 youth; however due to the innovations in training methodologies and overall project management strategies utilised by the consortium, there were over 6300 young people successfully completed their training courses.
- **Practical Sessions Including Teamwork**: This was a hands-on learning process where trainers could use and become comfortable with the various tools and equipment required to deliver the training to youth. The practical session complimented the practice classroom exercises; trainers were engaged in small group exercises in several practical demonstrations for the duration of the training (McEwan et al., 2017).
- **Illustrative Talks**: This is a lecture method accompanied by illustrations using training materials, including audio-visual aids, documentaries of success stories and case studies. Trainers were taught how to use audio-visual equipment and apps to enhance their training in the classroom.

TOT in CGJR

The experience of most graduates in Sierra Leone shows that having a college diploma and/or degree does not guarantee that they possess the competencies to be effective in the workplace nor in assuring them a job (Meeks, 2017; Eisner, 2010; Jackson, 2009). Employers look for people who bring hard and soft skills to the workplace. Employers are more eager for their workforce to possess more critical soft skills, including good communication skills, working with others, thinking critically and being more effective in managing their time (Meeks, 2017).

The counsellors trained in CGJR were instructed on how to use standard literature, training materials and methodology. The overall aim of the ToT was to enhance the skills and competencies of trainers to deliver training courses to boost youth employability by providing appropriate and viable education to meet job market demand and supply.

Counsellors on the completion of the training were able to:

Describe the philosophy, approaches and methodologies of the CGJR skills programme.

Explain the strategies for providing a safe learning environment to train young people in CGJR.

Deliver training using necessary tools to help students make informed educational and occupational choices to develop their Career Guidance and Job Readiness capacities.

Comprehend and facilitate the 15 modules in the training manual, including a module on Climate Change and Sustainable Development to youth within the academic year timeframe to deliver the certificate programmes.

Facilitate training programme to prepare young people to develop their soft employability skills including (a) problem-solving, (b) communication, (c) developing CVs and cover letters, (d) interviewing for a job, (d) work ethics, (f) teamwork, (g) decision-making, (h) building self-esteem, (i) behaviour change, (j) Climate Change and Sustainable Development and others.

The Table 5.1 summarises the 3-day CGJRS TOT programme, including the methods used to interact with and educate the trainers and how the content was spread over the days.

Reflection on the Effectiveness of the Training Methods

Approximately 8 months after the implementation of the TOT to training instructors and counsellors on the methods, content, tools and equipment to deliver the TVET training courses to youth in Sierra Leone, the consortium implemented a qualitative survey to assess the quality and effectiveness of the training delivered by instructors and counsellors. Emphasis was placed on evaluating the methodologies used to deliver the training courses. A summary of the stakeholders' views from the three components of the YEEP is below, utilising qualitative analytical tools (OECD, 2019; Clarke & Braun, 2013).

Component 1: Skills for Employment Programme

The feedback came from the M&E exercises conducted by the consortium and NAYCOM during the course's implementation and after executing the TOTs on the new content, knowledge and skills of curricula and the methodologies' effectiveness to deliver the new skills to youth.

A senior government officer commented on the effectiveness of these methodologies: 'In the area of effectiveness and relevance of the methods used in the project, it is important to note that the methodologies implemented are speaking to the core needs of the project' (1GN).

Another senior government officer supported the effectiveness of the methods used by the consortium: 'The consultative method is responsible for the establishment of the mutual understanding between implementors – PIs and the consortium. Experimentation is another of the successful methodologies implemented by the consortium. Throughout the project, there is evidence of constant testing of curricula and development strategies. This is supported by rigorous review and feedback of the performance of strategies between the consortium and the PIs' (2GN).

Senior lecturers/counsellors and head of PIs praised the effectiveness and variety of the range of methodologies utilised in the delivery of the course content: 'The programme utilises a range of viable methodologies and supporting training material, including demonstrations, field trips, flip charts and projector as aids to support training. Students have been able to comprehend the methods and content of the courses, and they have been remarkably effective' (1C1).

The youth commented on the effectiveness of the methodologies used by instructors in the training of the priority skills: 'The methodologies used are exceptionally good. These methods are a combination of demonstrations, notes, provision of practical exercises, and charts. The direct teaching of students in the classroom and lectures have been working well. At the same time, the provision of demonstrations without equipment and tools are not working well' (AC1).

All the respondents to the data collection survey praised the effectiveness and relevance of the wide range of methodologies used by instructors for the project's success. Respondents stated that the methods they were using during the training were acquired from the TOTs. These methods are responding to the core needs of the youth and the project. Some respondents specified the consultation's effectiveness and the importance of getting partner agencies to sign on to the MOU as the primary vehicle for building synergy and cooperation between the project implementors and the partner agencies. This strategy was an important

Table 5.1 CGJRS TOT programme timetable

Topics to be covered	Objectives	Summary of content	Methodologies	Indicators for monitoring
Day 1 topics: Career development Introduction to career development Career guidance Career planning Career development Job search What is job search and conducting an effective job search? Job search skills Using different media Job application process Ways to apply for a job Completing a detailed job application form Contacting the potential employer Job interviews Interview skills Different interview styles Mock interviews Guidelines for improving a candidate's chances Professional time management How to prioritise Time management skills and techniques How to effectively delegate Organising and setting SMART goals	Participants understand the YEEP programme and the career guidance and job readiness component of the project Participants understand the conceptual framework of the activities under the career guidance and job readiness programme and adapt the framework to the reality of their individual institutions Participants review their engagement and deliverables under the career guidance and job readiness programme as per the MOU Participants are capacitated with emerging information, methods and tools to help them deliver career planning and development training, job search skills, the job application process, job interviews and professional time management	Registration, introduction and icebreaker Orientation (statements and opening) Overview of the YEEP programme, goals, objectives and purpose of the training Training methodology Ground rules Completing application forms Writing application letters Writing cover letters Preparing CVs Preparing for job interviews Participating in mock interviews Presenting themselves in the proper appearance for getting the job Practising adhering strictly to time at work Reporting for work on time Planning their work at each time	Presentations Teaching explanations and instructions Motivational talks Discussions Exercises Roleplays	Attendance sheet Training evaluation forms Training report Radio discussion feedback
Day 2 topics: SWOT analysis What is SWOT analysis? When to use SWOT analysis SWOT analysis exercise – Templates Communication skills Definition and elements of communication The process and components of communication Communicating in work teams and non-verbal communication Communications etiquettes Communicating in business meetings and 10 top communication skills Work ethics Work ethics definition and discussion Why work ethic is important Work ethics that are in high demand How to develop a strong work ethic Time management skills and techniques Teamwork Definition and purpose of teamwork What makes the team successful? Team building Leadership skills What is leadership? Important leadership skills	Participants understand and can deliver training in SWOT analysis, communications skills, work ethics, teamwork leadership skills Participants are empowered with tools to support understanding, team and business communication, building a team and problem-solving	Recap What is SWOT analysis and SWOT analysis exercise –templates The process and components of communication Work ethics that are in high demand How to develop a strong work ethic	Discussions Group work Exercises Games	Attendance sheet Training evaluation forms Training report

(continued)

Table 5.1 (continued)

Topics to be covered	Objectives	Summary of content	Methodologies	Indicators for monitoring
Day 3 topics: Decision-making skills on building resilience to adapt to climate change and calamities Decision-making process Anatomy of a decision How to improve problem-solving and decision-making skills Creativity in decision-making process Building self-esteem What is self-esteem? Improving and building self-esteem Assertiveness and confidence Asking for and making what you want Behaviour change Introduction to behaviour change Behaviour management Cognitive behaviour modification	Participants understand and convey decision-making skills, building self-esteem, behaviour change, building community resilience for climate change Participants are capacitated with new information and tools and learning methodologies	Creativity in decision-making process Improving and building self-esteem Assertiveness and confidence Behaviour management Cognitive behaviour modification Climate change, impacts, vulnerability and adaptation Community-based climate change adaptation and resilience Assessing climate change vulnerability and capacity; planning for adaptation Session roundup	Presentations Discussions Exercises Assignments Motivational talks	Attendance sheet Training evaluation forms Training report

mechanism to engender respect and the recognition required to move the project forward. The MOU was also innovative because the Government of Sierra Leone recognised partner agencies as a valued collaborator in the national development and planning processes (Chap. 3).

The issues of experimentation were discussed as a credible methodology by respondents from the government. This involved the constant testing of curricula and development strategies throughout the project. Due to the rapid changes brought about by the impact of climate change and other development calamities such as the Ebola epidemic and the coronavirus pandemic, the consortium focused its efforts on the development and testing of curricula in the three skilled areas, namely, Sustainable Agriculture (Sustainable Aquaculture), Sustainable Hospitalities and Sustainable Construction. The aim of testing the new curricula was to build the resilience of youth and their communities to adapt to the impact of climate change and the emerging calamities. Supported by rigorous reviews and feedback among the consortium team, NAYCOM and the PIs, the curricula were revised and then approved by the NCTVA upon submission.

Instructors, counsellors and youth were excited about the innovative methodologies used in the training sessions. These methods include face-to-face-one-to-one training, lectures in groups and whole class exercises, written supporting materials championed by research via e-learning and focus group learning exercises (during the COVID-19 pandemic). These training methods were tailored to meet the needs of youth who completed formal education and those who went through non- and informal training, including those who completed the traditional school system and those who dropped out of formal education and decided to acquire a skill through NCTVA training course. Field trips and other demonstration activities were organised to support classroom training and learning activities.

Despite the effectiveness of the methodologies used by instructors and counsellors in the classroom, youth were concerned that the delay in the arrival of the tools created additional pressures on instructors to implement demonstration and practical activities (Olaniyan & Okemakinde, 2008; ILO, 2010; Mukhtara & Ahmadb, 2014; UNDP, 2015). Due to the COVID-19 pandemic outbreak, the tools and equipment were not delivered to PIs until July 2020. However, despite the lateness in the arrival of tools and equipment, over 310 youth graduated from employment component skills. Also, as a result of the pandemic, the project was extended to November 2020 to allow the youth to successfully complete the training.

Component 2: Career Guidance and Job Readiness Skills

A senior government official described the methodologies used in component 2 as one of experimentations throughout the project implementation cycle: 'The project is constantly putting new structures, capacities in place in entrepreneurial and economic activities. Here is where the experimentation becomes a constant methodological process practised throughout the project's life cycle' (2GN).

A senior counsellor highlighted the innovative methods used by trainers to deliver knowledge and skills to youth: 'Several innovative methods were used by trainers, including brainstorming, group work, role play, case studies and presentation and discussions. They were effective because they involve the active participation of participants in the course. Young people were engaged in the selection of the jobs they are interested in pursuing. This gives young people hope that all is not lost' (4C2).

Youth who participated in the training programmes highlighted the variations and quality of the methods used by lecturers to deliver the training: 'The lecturer usually uses the teaching methods such as role-playing, demonstrations, explaining, and they seem to be very effective. The quality of teaching from the staff was very high' (EC2).

Most senior officers of the government participating in the data collection exercise stated that the methods used by lecturers in delivering the Career Guidance and Job Readiness Skills training programme to youth were effective. A senior official stated that this was a situation where experimentation became a constant methodological process 'practised throughout the project's life-cycle'. The effectiveness of the methodology was because the consortium invested heavily in implementing two TOT workshops, resulting in doubling the number of counsellors trained in rural and urban universities, colleges and TVET centres. The consortium also developed a new curriculum, prepared training manuals and shared innovative training methodologies. The ToT workshops were held in Bo and Freetown, respectively. The original YEEP requested the consortium to train 30 counsellors from Freetown in Career Guidance and Job Readiness Skills; however, 60 counsellors were trained from Freetown, Bo, Kenema and Makeni in two 1-week intensive programmes. Experts in the field were hired to work alongside experts from the consortium delivering the programme.

All the youth attested to the quality of the methods used by counsellors to deliver the training course. One youth said: 'I love the methods used by the instructors because they made it easier to understand the key terms of the programme...the methods are very effective, and the variety of methods used by the instructors made it very easy for students to understand the information given' (CC2).

Another youth commented that 'the facilitators are well trained and skilled and have knowledge in the courses they are training us in... absolutely the methods are highly effective because the instructors are making the training very practical. The practical method used by counsellors are an advantage, and they are always present for the training' (AC2).

Youth were motivated and creative in using Facebook, WhatsApp and other apps to work with their colleagues to ensure the delivery of the course in rural communities, despite universities and colleges being closed during the COVID-19 pandemic. This has resulted in over 6200 youth successfully completed the certificate course in Career Guidance and Job Readiness Skills. The information provided is in line with the literature (ILO, 2010, Paschen, 2015; UNDP, 2015, Stochowiak et al., 2016).

Component 3: Graduate Entrepreneurship Programme

The consortium invested a considerable amount of time and resources in implementing TOTs for instructors and managers of the business service centre (BSCs) responsible for implementing the Graduate Entrepreneurship Programme (GEP). Initially, the consortium was to work with one BSC in Freetown, the capital of Sierra Leone. However, besides the centre in Freetown and without any additional financial resources, the decision was taken to work with two additional BSCs in the rural provinces of Kenema and Makeni.

A senior manager of a BSC expressed his views on the effectiveness of the methodologies used in these centres: 'Some of the training methodologies includes but not limited to Slides, Icebreakers, flip charts, group work and presentations, individual assignments and presentations, video documentaries, case studies, experience sharing, motivational insights, field exercises, demonstration, simulation, focus group work and discussion, pictorial analyses and extract hard copies of the training manual were used to facilitate the training process. The training methods were relevant and effective as the choice of methodologies adopted in each training session depending on the lesson's contents. Therefore, mixed forms of methodologies were used for the training. This implies that the methodologies were instructive and support active class participatory approaches' (2C3).

The youth were thrilled with the range, diversity and effectiveness of the methodologies used by the business trainer/instructors. Youth (BC3) posited: 'The methods delivered by the instructors are very much in place to ignite my understanding and the understanding of all the other trainees, I believe. Sure, the methods are effective because everything is based on reflection and practice' (BC3).

All the managers, instructors and youth participating in the data collection exercise of component 3 praised the wide variety and effectiveness of the methods used by the instructors to deliver GEP. Another BSC senior manager and Instructor commented that a considerable amount of consideration went into the selection of the methodologies utilised in the programmes: 'The method used to train graduates were appropriate as it gave participants the free will to engage themselves in friendly debates and helped them identify answers to some of their questions; it also helps them to understand the topics easily especially during group works' (1C3). This is testimony to the considerable amount of time and resources invested in implementing the TOTs for instructors and trainers.

The Freetown consortium team made efforts to ensure that the methods selected for the training of the instructors were effective and based on the needs of youth.

They use lots of technology to deliver the training, and the atmosphere was very relaxing. Yes, the methods were very effective; they take their time to explain and ensure we understand. They also give us the opportunity to make presentations to boost our confidence levels. (A youth – GC3).

A senior manager of the government endorsed the appropriateness and effectiveness of the diversity of the methodologies used by PIs: 'I have been to the field, and the methodologies that are in place in the PIs are appropriate…the methodologies are well thought through, and they are working very well' (1GN).

Well-staffed and adequately funded training institutions are essential to skills development strategies and policies. Periodic reviews must be implemented to assess the effectiveness of the skills training initiative in meeting their goals of addressing the gaps and mismatch between the skills provided by training and educational institutions and that which is required by the market. TOTs promote efficiency in using scarce resources (Carter, 2008; ILO, 2010; UNDP, 2015).

Summary of the Chapter

In summary, the chapter is split into two segments. The first segment explored the methodologies used by the consortium to collect data from stakeholders on the gaps and challenges regarding the curricula review process. It utilised the information provided from the data collection process to inform the strategies for reviewing the curricula and the development of the training manuals for the curriculum processes. The approved curricula include (a)

Sustainable Hospitalities (Tourism and Hotel Management), (b) Sustainable Agriculture (with emphasis on Aquaculture), (c) Sustainable Construction (Carpentry, Masonry and Fabrication), (d) Career Guidance and Job Readiness Skills and (e) Entrepreneurship Skills development. The consortium implemented five TOTs to train instructors and counsellors to deliver the TVET curricula to youth. Multiple methodologies were used to collect data on YEEP components: interviews with key informants, including senior government officials and PI heads and instructors involved in the implementation and FGDs among groups of young people from rural and urban areas for the three components, (a) Skills for Employment training (component 1), (b) the Career Guidance and Job Readiness training (component 2), and (c) the Graduate Entrepreneurship Programmes (Component 3).

This chapter also documented the strategies employed by the NCTVA to implement the curricula approval process. The COVID-19 pandemic national lockdowns caused colleges and universities to close their operations which affected the approval of the curricula by NCTVA as the curricula committee meeting had to be postponed. It is important to note that this delay did not stop the approval of the curricula. Youth and instructors of PIs were so motivated that they continued to deliver the course through innovative e-learning processes until the project was completed in November 2020.

The second section discussed the methodologies and implementation of the Training of Trainers programmes. This resulted in enhancing the institutional capacity of the PIs to deliver the training to youth. The chapter explored the methodologies and strategies utilised by the consortium to train instructors, lecturers and counsellors to use the curricula and training modules to deliver the training courses to youth in their respective institutions. The chapter concluded by presenting a qualitative analysis of stakeholder views regarding the effectiveness and relevance of the methodologies used by the training instructors.

References

Angelsen, A., Larsen, H. O., Lund, J. F., Smith-Hall, C., & Wunder, S. (2011). *Measuring livelihoods and environmental dependence; methods for research and fieldwork*. Earthscan.

Baxter, P., & Jack, S. (2008). Qualitative case study methodology: Study design and implementation for novice researchers. *The Qualitative Report: McMaster University, West Hamilton, Ontario, Canada, 13*(4), 544–559.

Carter, R. (2008). *Caribbean youth: An integrated literature review*. University of Hawaii Press.

Clarke, V., & Braun, V. (2013). Teaching thematic analysis: Overcoming challenges and developing strategies for effective learning. *The Psychologist, 26*(2), 120–123.

Creswell, J. W. (2009). *Research design, qualitative, quantitative and mixed methods approaches* (3rd ed.). Sage.

Earl, M. J. (2001). IT strategy in the new economy. In J. Pickford (Ed.), *Mastering management 2.0: Your single-source guide to becoming a master of management, financial times*. Prentice Hall Publishing.

Eisner, S. (2010). Grave new world? Workplace skills for today's college graduates. *American Journal of Business Education, 3*(9), 27–36. https://doi.org/10.19030/ajbe.v3i9.478

Government of Sierra Leone. (2019). The Government of Sierra Leone's new Medium-term National Development Plan (MTNDP) 2019–2023, Freetown.

Guest, G., MacQueen, K. M., & Namey, E. E. (2012). *Applied thematic analysis*. Sage.

Hernandez-Delgado, E. A. (2015). *The emerging threats of climate change on tropical coastal ecosystem service, public health, local economies and livelihoods sustainability of small islands: Cumulative impacts and synergies*. Marine Pollution Bulletin, University of Puerto Rico, Centre for Applies Tropical Ecology and Conservation, Coral Reef Research Group Puerto Rico, USA.

Hickman, S. (2002). *Handout on Strategic Information System Planning (SISP) – Information Technology and Business, Masters in Business Administration*. University of Greenwich, London.

ILO. (2010, November). *A skilled workforce for sustainable growth: A G20 training strategy*. International Labour Office.

IMF. (2019). Sierra Leone: Economic Development Documents-National Development Plan 2019–2023: IMF Staff Country Reports, International Monetary Fund Africa Dept. Country Report No. 19/218.

IPCC. (2014a). Climate Change (2014): Synthesis report. In *Contribution of Working Groups I, II and III to the fifth assessment report of the Intergovernmental Panel on Climate Change* (p. 151). IPCC.

IPCC. (2014b). Summary for policymakers. In: Climate change impacts, adaptation and vulnerability. In C. B. Field, V. R. Barros, D. Dokken, K. J. Mach, M. D.

Mastrandrea, T. E. Bilir, M. Chatterjee, K. L. Ebi, & Estrada (Eds.), *Contributions of the Working Group II to the Fifth Assessment Report* (pp. 1–32). Cambridge University Press.

Jackson, D. (2009). An international profile of industry-relevant competencies and skills gap in modern graduates. *International Journal of Management Education, 8*(3), 29–58. https://doi.org/10.3794/ijme.83.288

McEwan, D., Ruissen, G. R., Eys, M. A., Zumbo, B. D., & Beauchamp, M. R. (2017). The effectiveness of teamwork training on teamwork behaviors and team performance: A systematic review and meta-analysis of controlled interventions. *PLoS One, 12*(1), e0169604. https://doi.org/10.1371/journal.pone.0169604

Meeks, G. A. (2017). *Critical soft skills to achieve success in the workplace: Walden University Scholar Works, doctoral study submitted in partial.* Walden University.

Mukhtara, M. I., & Ahmadb, J. (2014). Assessment for learning: Practice in TVET: 4th World Congress on Technical and Vocational Education and Training (WoCTVET), 5th–6th November 2014, Malaysia: Elsevier. *Procedia – Social and Behavioral Sciences, 204*(2015), 119–126.

Mukhtara, M. I., & Ahmadb, J. (2015). Assessment for learning: Practice in TVET. *Procedia – Social and Behavioral Sciences.* https://doi.org/10.1016/1.sbspro.2015.08.124

Musau Zipporah. (2017). Reducing inequalities: Africa grapples with huge disparities in education: Higher enrolment numbers mask exclusion and inefficiencies: From Africa Renewal: December 2017 – March 2018.

OECD. (2013). *The nature of policy change and implementation: A review of different theoretical approaches: Organisation for Economic Cooperation and Development.*

OECD, DAC. (2019). *Better criteria for better evaluation revised evaluation criteria definitions and principles for use OECD/DAC network on development evaluation, better policies for better lives series.* http://www.oecd.org/dac/evaluation/revised-evaluation-criteria-dec-2019.pdf

Olaniyan, D. A., & Okemakinde, T. (2008). Human capital theory: Implementation for education and development. *Pakistan Journal of Social Development, 5*(5), 479–483.

Poschen, P. (2015). *Decent work, green jobs and the sustainable economy: Solution for climate change development.* International Labour Organisation.

Pratchett, M. S., Wilson, S. K., Graham, A. J., Munday, P. L., Jones, G. P., & Polunin, N. V. C. (2009). Coral bleaching and consequences for motile reef organism; past, present and uncertain future effects. In M. J. H. Van Oppen, J. M. Lough (Eds.), *Coral bleaching: patterns, processes, cause and consequences ecologic studies* (vol 205, pp. 139–158). Spring- Verlag, Berlin.

Robson, C. (2002). *Real-world research, a resource for social scientists and practitioner researchers.* Blackwell Publishers.

Saunders, M., Lewis, P., & Thornhill, A. (2000). *Research methods for business* (2nd ed.). Pearson Education Limited.

Simmons, A. (2021). *Impact of climate change on youth in small island communities Vol 1* (Sustainable development goals series, climate change). Springer Nature.

Simona, M., Elitsa, G., & Alma, S. (2014). *Assessment study of technical and vocational education and Training (TVET) in Myanmar; ILO DWT for East and South East Asia and the Pacific.* ILO 2014.

Stochowiak, S., Robles, L., Habtermanion, E., & Maltry, M. (2016). *Beyond win: Pathway for policy implementation.* ORS Impact/The Atlas Learning Project.

Sustainable Development Solutions Network – Youth. (2020). *Youth solution report (4th edition) sustainable development solutions network – Youth, UNDP, New York.* Available at: https://www.undp.org/content/undp/en/home/librarypage/democratic-governance/youth-solutions-report-2020.html. Last accessed April 2021.

UNDP. (2015). *Human development report 2015.* United Nations Development Programme.

UNDP. (2020). *Human development report 2020.* United Nations Development Programme.

UNEP. (2021). *GEO-6 for youth.* UN Environment Programme.

UNESCO. (2010). *Climate change education for sustainable development: The UNESCO climate change initiative, Paris.*

UNESCO. (2014). *Shaping the future we want: UN Decade of Education for Sustainable Development (2004–2014).* DESD Monitoring and Evaluation.

United Nations. (1987). Our common future: Report of the World Commission on Environment and Development: Our Common Future Transmitted to the General Assembly as an Annex to document A/42/427 – Development and International Cooperation: Environment, New York.

Williams, D. (2013). *Redesigning Schooling – 3: Principled curriculum design: Redesigning School, the Campaign schoolish vision for education.* SSAT (The School Network) LTD.

Yin, R. K. (2003). *Case study research: Design and methods* (3rd ed.). Sage.

Building the Resilience of Youth in Sustainable Livelihoods Through Entrepreneurship in West Africa

6

Abstract

This chapter summarises the outcomes of the YEEP implementation process and deliverables, and proposed actions/recommendations for enhancing the overall implementation of the project in the future. It also presents the lesson learned and best practices to enhance tertiary and TVET institutions involved in the YEEP's management and delivery. Overall, these recommendations are critical to building resilience among youth in LCDs in West Africa to adapt to climate change and other development calamities in the future. The chapter analyses the implementation of the conceptual framework in terms of the challenges and gaps in its implementation and highlighted the need for future research to enhance the body of knowledge on youth, climate change adaptation and sustainable livelihoods, specifically in skills training for employment through entrepreneurship.

The chapter explores the effectiveness and challenges of implementing top-down and bottom-up approaches to deliver skills for sustainable livelihoods and discusses the need for youth to build resilience to adapt to climate change and development calamities such as Ebola and COVID19. The chapter concludes by summarising the relevance of the strategic tools such as cultural web, IT/IS frameworks and business process re-engineering as mechanisms for further assessing the gaps and weaknesses of institutions and mechanisms with responsibility for implementing Climate Change education for sustainable development and recommends specific actions to be taken to enhance the employability and development of youth in the future.

Keywords

Entrepreneurship development · Climate change adaptation · Employment · Sustainable development · Resilience building · Cultural web · Youth development · Top-down approach · Bottom-up approach · Education for development · Poverty alleviation

Introduction

The Government of Sierra Leone and development agencies are fully aware of the severity of the unemployment situation facing young people and how climate change and other development challenges exacerbate these issues. The government's human capital development approach addresses issues relating to youth development in Sierra Leone (IMF, 2019; Government of Sierra Leone, 2019; UNDP, 2019). Youth development is viewed by the government as a viable approach to enhance human capital to transit the economy

A. Simmons, *Climate Change Adaptation Framework and Youth Entrepreneurship in West Africa*, Sustainable Development Goals Series, https://doi.org/10.1007/978-3-030-85754-7_6

to a middle-income economy by 2035 and reduce the impact of poverty across the country. It is important to note that seven of the eight development clusters in the National Development Plan are linked directly to youth development in Sierra Leone. Although youth development is an individual cluster identified in the National Development Plan, all other areas tackle youth development issues.

The chapter summarises and discusses the YEEP outcomes, the lesson learned and best practices to enhance NAYCOM and PIs' capacity in managing and delivering the project. The chapter explores the effectiveness and challenges of implementing top-down and bottom-up approaches to provide entrepreneurship skills for sustainable livelihoods and discusses the need for youth to build resilience to adapt to climate change and development calamities such as Ebola and COVID-19. The chapter concludes by summarising the relevance of strategic tools used during the project, such as the cultural web, to assess the gaps and weaknesses of NAYCOM and recommends specific actions to enhance the employability of youth in Sierra Leone and Africa.

Scaling Up the Youth Employment and Entrepreneurship Project in Sierra Leone, West Africa

The Government of Sierra Leone recognises the significance of YEEP as the vehicle to address the severe employment challenges facing youth and enhance their capacity to contribute to national development. The table above shows that the government approach to tackling young people's challenges is precisely correct; however, these approaches need to be supported by investments in human capital development and resilience building to ensure that youth employment and development goals become a reality (Table 6.1).

Table 6.1 Clusters relating to youth development as indicated in the National Development Plan, Government of Sierra Leone

Development clusters	Elements of the clusters relevant to youth development
Human capital development	Strengthening tertiary and higher education
	Ensuring free quality primary and senior secondary education
	Accelerating health-care delivery
Diversifying the economy and promoting growth	Improving productivity and commercialisation of the agricultural sector
	Improving productivity and sustainable management of fisheries and the marine sector
	Revitalising the tourism sector
	Promoting an inclusive rural economy
Infrastructure and economic competitiveness	Improving information and communication technologies
	Fostering private sector growth
Governance and accountability for results	Fighting corruption and illicit financial flows
	Strengthening decentralisation, local governance and rural development
	Building public trust in state institutions
Empowering women, children, adolescents and persons with disability	Increasing investment in children and adolescents
	Empowering women (a large percentage are youth)
Youth employment, sports and migration	Increasing youth employment and empowerment, revitalising sporting activities across the country and addressing migration challenges
Addressing vulnerabilities and building resilience	Building national environmental resilience
	Strengthening forest management and wetland conservation
	Improving disaster management governance

Government of Sierra Leone (2019) and IMF (2019)

YEEP is relevant to address the development needs of youth Sierra Leone, having doubled and tripled all proposed deliverables and outcomes due to NAYCOM becoming a stronger and more viable entity from its investment in enhancing its institutional capacity. The consortium instituted the cultural web, IT/IS systems and business process reengineering to assess the culture, governance and decision-making processes within the

NAYCOM regarding its readiness to implement changes to enhance its institutional capacity and move into a sustainable development direction. Upon completing these institutional assessment tools, time did not permit the consortium team to work with NAYCOM to ensure that the governmental agency implemented the strategies. This action was not part of the deliverables of the original project.

This chapter is divided into three sections: (1) a summary of the project's objectives and outcomes, (2) the lessons learned and possible best practices of the project and (3) recommendations for scaling up the YEEP project.

Institutional Capacity-Building Tools

The strategies for enhancing the capacity development of NAYCOM were assessed internally and externally. Internally, the cultural web, IT/IS systems, business process reengineering and critical path was used to evaluate the international capacity of the organisation to support sustainable change management. From an external perspective, investment in technology as a mechanism for ensuring that the organisation benefits more significant numbers of youth across Sierra Leone, including e-learning, Internet and websites business development through entrepreneurship, was assessed.

Application of the Cultural Web

NAYCOM's role culture is typical for a public service entity in Sierra Leone (Handy, 1986). Although the culture in NAYCOM is not as restrictive compared to other government organisations, such culture makes it extremely difficult to implement any changes to enhance the efficiency and effectiveness of the organisation. Through consultations with senior management and other members of staff, the consortium developed an alternative culture for NAYCOM to boost its efficiency and effectiveness and enhance its human capacity to achieve the government's sustainable development goals articulated in the National Development Plan, shown in Chap. 4 (IMF, 2019; Government of Sierra Leone, 2019). The consortium could not work with the Commissioner of NAYCOM to implement the change management cultural model due to the outbreak of the COVID-19 pandemic and change management not being included in the YEEP deliverables. The consortium management advised the team in Freetown to expend its resources on attaining the project deliverables rather than out-of-scope activities.

It is important to note that a shift in the culture of the National Youth Commission is required for the organisation to achieve its policy outcomes and strategic goals (Slack et al., 1998; Johnson and Scholes, 2002). Given organisational complexities and the lack of confidence of some civil servants in the institution, there may require the implementation of broader change management across all governmental institutions driven by the political head of the country or his designate to demonstrate the commitment of the government to the change process and build the confidence of employees and the public to trust the process (Bernick, 2001; IMF, 2019; Government of Sierra Leone, 2019). As this was not included in the deliverable of the YEEP, the time did not permit the analysis of the adequacy of the resources (financial, human and material) available in the national budget to implement the change process within a given timeframe.

IT/IS System in Decision-Making at Strategic Levels

NAYCOM, like many other pseudo-public sector agencies in LDCs, modernises its governance, management and operations mechanisms through IT systems and Internet use. IT/IS was utilised to review how IT is used to support management to make strategic decisions to boost the overall operationalisation of the project and the institution. Although NAYCOM possesses an established IT desk and IT manager/officer, there is the absence of an IT strategy to support the organisation to achieve its policy goals and outcomes (Earl, 2001). There is also no programme to raise officials' awareness on how to apply ICT at the strategic level to give the agency a competitive advantage. While the consortium has developed a comprehensive programme to improve utilisation and bolster the existing IT systems, it was not implemented as it was out of scope.

Business Process Reengineering

The Government of Sierra Leone is mobilising resources from bi- and multilateral agencies to implement socioeconomic and governance projects across the country to support the achievement of the sustainable development goals as articulated in the National Development Plan (Government of Sierra Leone, 2019). Project drift is a severe development challenge affecting the implementation of projects in LDCs in Africa, with a high percentage of projects implemented by donor agencies in West Africa running behind implementation schedule (Niyonkuru, 2016). However, the challenges presented by the COVID-19 pandemic resulted in PI closures and the temporary abandonment of the project between April and June 2020. BPR and critical path tools were used to enhance the effectiveness and efficiency of the project management systems, leading to the successful completion of the project 2 months after its designated termination date. These analytical tools provided the factors that support or impede change and efficiency in the organisation's project management process (Chap. 4).

Investing in Technology to Enhance Climate Change Adaptation

Building national environmental resilience, improving disaster management governance and strengthening tertiary and higher education are enshrined as strategies in the Sierra Leone National Development Plan (IMF, 2019). The Government of Sierra Leone perceives that these strategies should be driven through a human capital development approach (IMF, 2019). Therefore, for this strategy to become a reality, the government must ensure increases in investment in technology to build the resilience of youth to adapt to the impact of climate change and other development calamities. Youth have a clear understanding of what needs to be done in terms of the technology to achieve sustainable livelihoods. When colleges and universities and other institutions were forced to close due to the national COVID-19 lockdown, youth used their phones and different innovative strategies to connect to PIs and continue their training despite none of these institutions having a functional e-learning platform to provide such training. The training continued because a dedicated group of trainers, counsellors and administrators from PIs committed themselves to provide training in skills for employment to empower the youth to acquire appropriate skills and knowledge for the future. Over 60% of youth who successfully completed the project were young women. NAYCOM is committed to creating the enabling environment within the Sierra Leonean society where women can make empowered choices; an educated, professionally active mother can inspire and stimulate her children (UNDP, 2015).

Technology can play a critical role in building youth resilience in LDCs to adapt to climate change. Therefore, technological innovation should be viewed in its broadest form comprising development, implementation and production of new products, procedures and processes to meet market needs. The strategy should focus on enhancing the community's adaptive capacity through a human capacity development approach to help people manage climate change uncertainties such as rising temperature, sea level rise and natural disasters, biodiversity reduction, deforestation and increased variety and severity of diseases. It should also provide them with the tools to address these climatic changes and enhance sustainable livelihoods and overall development (Nurse et al., 2014, Hernandez-Delgado, 2015).

One of the project outcomes was developing an online digital platform utilising the Internet to provide opportunities for youth entrepreneurs to collaborate with business mentors to provide mentorship support and exchange of new knowledge and skills to the youth entrepreneurs. Despite the challenges with Internet connectivity across Sierra Leone and the low-wave bandwidth, there is still a future for ICT to drive the technological revolution in Sierra Leone with investment in technology. The government and people must perceive technological innovation as a creative tool to empower youth with knowledge and skills to build their capacity to adapt to climate change and the various calamities that confront their daily lives. It also must be considered a vehicle to assist rural and urban youths in developing sustainable livelihood projects that possess the capacity to create viable employment and enhance the country's economic and social development (Poschen, 2015; UNDP, 2015). Youth adapt more quickly to the use of technology than an older person. Therefore, the government should invest in technology, particularly Internet bandwidth to enhance capacity, provide e-learning support to tertiary and TVET institutions and modernise training tools and equipment of the institutions involved in employment skills programmes. It should also provide early warning systems to inform communities of approaching climatic weather systems, overflowing rivers and other climate-related calamities.

The consortium was impressed with the quality of the sustainable livelihood projects originating from the business pitching initiative. It is believed that if the government continues to invest in new technical training programmes and equipment to empower the youth to a) develop green energy initiatives at community and national levels; b) continue to adapt to landslides, temperature rise and other climate stressors; c) create drought- and disease-resistant crops to enhance food security and d) reduce land degradation and biodiversity loss through ethical ecosystem management; this investment may enhance the trust and pride of youth and contribute to stemming the avalanche of youth fleeing the region in search of greener pastures in Europe and the Middle East (UNDP, 2015; Poschen, 2015). Although technologies are essential for driving knowledge, skills development and sustainable livelihoods, the 'technological revolution has been accompanied with rising inequalities' (UNDP, 2015, p100). Although the new world of work on human development is promising, these promises are not yet fulfilled. The reality is that rapid technology expansion without national training programmes will leave many citizens behind (UNDP, 2015).

Implementation of the Conceptual Framework on Climate Change Adaptation for Youth in Africa

The conceptual framework on climate change adaptation for youth was developed for SIDS as a mechanism to support research work on the impacts of climate change on youth in small island communities. In the absence of an appropriate conceptual framework designed to explore the complexities between CCA and sustainable development for youth and their communities in Africa, the framework developed for youth in the Caribbean SIDS was piloted successfully in Sierra Leone by NAYCOM, funded by AfDB. It is hoped that that work will be a tangible and sustainable model on which research on adaptation for entrepreneurship will occur among youth in West Africa. The framework can contribute to enhancing the body of knowledge on climate change adaptation and youth development. It will provide policymakers and practitioners in Africa with practical tools and strategies for developing and implementing Climate Change Education and adaptation strategies to prepare young people to build their resilience to adapt to the impact of climate change through entrepreneurship development (ILO, 2010; Simmons, 2021). M&E is included as an essential feature of the framework to provide feedback on the performance of the various stages of the framework and strategic advice to move the framework forward (Harley & Minnen, 2010).

The Government of Sierra Leone has expended considerable amounts of technical and financial resources through the Anti-corruption Agency to ensure that it can deal with the crippling impact of corruption in government ministries and departments across the country. Although corruption was not a factor observed during YEEP's implementation, due to the devastation that corruption can cause, it is pertinent to include anti-corruption indicators in the new conceptual framework when working in West Africa. The government must be congratulated for providing leadership in Africa through implementing a mammoth anti-corruption campaign.

Like other West African LDCs, Sierra Leone is exposed to numerous development calamities that have devastating impacts on human capacity-building projects. In 2014–2015, the region was impacted severely by the Ebola epidemic, resulting in the deaths of thousands of people. In 2018, the country suffered fatalities when landslides caused by deforestation, intensive rainfalls and poor land management practices buried a community adjacent to Freetown (Hernandez-Delgado, 2015). Most recently, as of completing YEEP's pilot, the COVID-19 pandemic impacted the project severely, resulting in the national lockdown and closures of universities, colleges and TVET centres. Development calamities must be included in the new conceptual framework due to the devastating impact they can have on development projects like YEEP and on the country as a whole (Angelsen et al., 2011; Pereznieto et al., 2011; Hermandez-Delgado, 2015; IPCC, 2014a, b, c).

Stakeholders' Perspective on the YEEP: Reasons for Participating in the Project

All stakeholders state that their involvement in the project was for varying reasons. The consultants were pleased with the high levels of enthusiasm by the government and PI officers. The government officials and PI instructors were motivated to participate in the project because they want to contribute to youth development in Sierra Leone; none said they were involved in the project because it was their job or the financial resources they could receive. Although YEEP provided additional responsibilities for these officials to their other duties in their post of employment, they persisted to ensure that the project deliverables were successfully implemented because of the capacities of the project to resolve the youth unemployment challenges of the country (Government of Sierra Leone, 2019; IMF, 2019). They participated actively in the YEEP because they recognised the project as a development programme that will enable youth to gain requisite skills and knowledge to gain employment and opportunities to contribute to the realisation of national development goals (Government of Sierra Leone, 2019; IMF, 2019). The representatives of partner institutions were enthused to participate in the project because signing the MOU with the government is a testament that the government values their contribution to achieving the national development goals.

YEEP on Youth Employment and National Development

To resolve the skills and knowledge mismatch in the economy, ToTs to train instructors, counsellors, and heads on the new content of the curricula and methodologies for delivering the three components of YEEP to youths were implemented (UNDP 2015, Carter, 2008). All respondents viewed the training methodologies as highly effective and relevant to the emerging development needs of skills development for employment training in Sierra Leone. The students praised the range and variety of methods used by instructors for the project's success, which provides evidence that the training in the TOTs was phenomenally successful. The YEEP pilot's success and eventual upscaling will be part of a growing library of development projects fostering and developing the emerging green economy (Tukundane et al., 2015; Kanyandago, 2010).

According to survey respondents, these methods respond to the core needs of youths and are relevant to the development challenges facing them and their communities. The respondents praised the effectiveness of the consultative process implemented by the consultant team. As there are high levels of mistrust between all sectors within the Sierra Leonean society, MOUs were signed between senior officials of partner agencies and the government to act as proof of partnerships, representing a shared commitment to the project. The MOUs were the primary vehicle for building synergy and cooperation between the project implementers, the government (NAYCOM) and the partner agencies. The MoU was also innovative because it recognised partner agencies by the Government of Sierra Leone as a valued and trusted partner in the national development and planning processes.

All stakeholders singled out the experimentation methodology as a unique methodology that empowered young people, building their self-esteem and overall confidence. The consortium team was involved in continual testing and assessment of curricula and development strategies throughout the project's life cycle to ensure its relevance to national development plans and its applicability to different locations and youth demographics. With the inclusion of climate change education and relevant hard and soft skills development, the new curricula will enhance youths' resilience to adapt to the impact of climate change and emerging calamities. It will bolster employability and proactivity, satisfying future market demands through employment and entrepreneurship opportunities to shape more sustainable and ethical business cultures.

Strategies for Upscaling YEEP

All the stakeholders who participated in the data collection process support the upscaling of the project. PI representatives agreed that the government should take the lead in coordinating and facilitating the project upscaling process (Matland, 1995; OECD, 2013; Schleussner et al., 2016). AfDB has committed to providing funding for upscaling YEEP after its pilot's successful completion. To support the upscaling, NAYCOM should put necessary bottom-up measures to allow PIs to participate more closely throughout the project proposal development, and an appraisal committee should be instituted to coordinate the work on the ground (Matland, 1995; OECD, 2013; Jaja et al., 2016; Stachowiak et al., 2016). AfDB's activities should include mobilising financial and technical resources to fund YEEP's upscaling of the project by putting mechanisms in place for staff and young people of PIs to participate in the project upscaling and implementation process. The key outcome of the new YEEP should be to provide opportunities for youth to gain decent employment upon their graduation from colleges, TVET centre and universities (ILO, 2010; Poschen, 2015).

The new project should follow the pilot project by providing financial support to PIs to purchase tools and equipment to support the new curricula and training programmes. The YEEP upscale project needs to adopt the new conceptual framework's approaches and methodologies and other tested innovations to deliver skills and knowledge to youths. The development of new skills for the employment curricula, procuring tools and equipment for training youth and working with PI to secure the latest technologies to boost IT/IS systems to assist with decision-making and the delivery of new knowledge and skills through e-learning should be included in the upscaled project (Thuraya, 2015; ILO, 2010).

The hybrid approach must be viewed as a vital element of the youth development process as it provides opportunities for PIs to work directly with government agencies on issues relating to youth development, employment, entrepreneurship and climate change adaptation (Stachowiak et al., 2016; OECD, 2013; Matland, 1995). It also provides opportunities to implement the climate change education for sustainable development models in Africa successfully (UNESCO, 2014, 2015; UNDP, 2015). To ensure the successful development and implementation of the new YEEP, NAYCOM should involve the key experts of the consortium to participate in the process leading up to the development of the project to reduce the duplication of efforts and wasting valuable limited time and resources.

Lessons Learned

Learning from previous projects is vital to the active project manager to continuously strive to become more effective at managing development projects (Vignos, 2014). This section provides project practitioners and managers with information on the lessons learned, practices evolving from the pilot and strategies to enhance existing practice in the field when upscaling YEEP to the national level. Lessons learned is defined as 'key project experiences which have a certain general business relevance for future projects…have been validated by a project team and represent a consensus on a key insight that should be considered in future projects' (Schindler & Eppler, 2003 p. 220).

'Making good decisions is a crucial skill at every level' (Drucker, 2006 p.120). It is a costly risk to the organisation to continuously repeat poor business decisions that it has made in the past. Therefore, methods to assess the current mechanisms, such as BPR, are pivotal in identifying underlying paradigms which hinder growth opportunities and create programmes to improve how such organisations run (Vignos, 2014).

Component 1

In component 1 (Skills for Employment), the delay of the tools and equipment and the installation in the PIs contributed immensely to slowing the learning process during the delivery of the skills for employment training programmes. These tools, purchased in China, were severely delayed because of the COVID-19 pandemic. In the future, NAYCOM and the AfDB should ensure that procurement processes are implemented 9 months to a year to prevent delays in the project implementation process. Project managers should also ensure that tools and equipment arrive approximately 6 months in advance of the new school year to make sure that tools and equipment installation and TOTs are completed before the project commences, and the process of training students is more streamlined (Vignos, 2014).

The curricula were submitted to NCTVA approximately 6 months before the start of the school year. However, it took over 9 months for the agency to complete the approval process. NCTVA complained of the limited technical personnel available and the lack of necessary financial resources to organise the consultative meetings and provide a stipend to experts to ensure quality feedback on the curricula; the onus was placed on the consortium to mobilise the financial resources. The late approval of the stages of the project by NAYCOM caused delays in the consortium obtaining the funds to support NCTVA with the curriculum review. This problem severely impacted negatively on the consortium's funding PIs to implement project activities. To address the delays by NAYCOM PIU regarding payment approvals for every completed stage, the project lead organised a series of meetings with the NCTVA director and PIs to implement the curricula during the approval process. A comprehensive M&E system was instituted which gave the benefit of real-time testing of the curricula and better familiarisation for PIs and NCTVA on running the education programme. Based on the feedback by implementing agencies, the curriculum and training manuals were amended and resubmitted to NCTVA mid-way through the training process for approval.

During the implementation of the upscaled programme, there must be a concerted effort on the part of NAYCOM and the implementing agencies to ensure that clear indicators and timelines are set for timely approval of the stages of the project to ensure that resources – financial, human and material – are available to support the smooth implementation of each project stage.

Component 2

In component 2, the Career Guidance and Job Readiness Programme, all training was temporarily suspended due to the outbreak of the COVID-19 pandemic and the national lockdown. The consortium implemented a series of consultations with PIs to shift to e-learning as a strategy for engaging youths. The move to e-learning was met with severe institutional challenges, including (a) poor network and data bandwidth, causing slow and disruptive Internet connectivity, (b) a lack of viable IT/IS systems supportive of e-learning in universities and colleges and (c) the absence of dedicated technology facilities in colleges (Thuraya, 2015; UNDP, 2015). While most students had no access to laptops or tablets to connect to the Internet, they had access to phones.

Working with PIs, the consortium devised innovative approaches using social media and virtual conferencing apps to share information with students, enabling meaningful dialogues with PIs, counsellors and instructors and their peers. The modules and other reading materials were uploaded to university websites and hyperlinked on social platforms to ease access. These approaches compensated for lost credit hours during the COVID-19 period. The consortium did not provide any financial resources to support these activities. In the future, the project budget must include contingency funds to respond to such events in an unpredictable climate.

Component 3

As evident in the other components, component 3 encountered resourcing constraints due to the late approval by NAYCOM and an unwillingness on the part of the consortium to pre-fund these gaps, impacting the smooth implementation of the activities in PIs. Although some PIs were able to pre-fund some activities upfront, most were unable to do so, resulting in the temporary suspension of project activities. Prolonged delays regarding the approval of the stages of the project deliverables must be avoided in future projects. Implementing agencies and government agencies involved with project implementation must develop an agreed timetable with milestones, including deadlines for stage completion, document submission and funding.

Young people are the drivers and leaders of change in any country; however, the quantity of financial resources available for sustainable finance youth development project is usually minimal. Governments and development institutions must create an enabling environment to empower young people through investments to avoid the development pitfalls that fragile states like Sierra Leone experience. The investment must focus on equipping youths to drive entrepreneurship development and growth to achieve national sustainable development (ILO, 2010; Poschen, 2015). The government should create a national revolving youth development fund to support entrepreneurship development projects run by young people. The projects evolving from the fund should be supported by investment in technology as a mechanism to maximise business capacity; increase innovation; modernise procurement, production and supply chain pipelines; and access local, regional and global markets (Prahalad, 2006; Hamel, 2001).

Best Practice of the YEEP

The Youth Climate Change Adaptation Framework

The Youth Climate Change Adaptation Framework is a testing model for implementing programmes to build youth resilience to adapt to climate change and enhance sustainable livelihoods. Initially developed for SIDS, amendments were made to make it relevant to African LDCs during the YEEP pilot, including:

- An M&E framework to provide feedback on the performance of individual stages.
- Broadening the development calamities to include Ebola and COVID-19.
- Anti-corruption as a development indicator; the Government of Sierra Leone is investing massively to fight corruption within public sector agencies (Fig. 6.1).

Adjustments were made to the revised conceptual framework on the adaptation of youth to new climate conditions to create the new Climate Change Adaptation Framework for youth in entrepreneurship in Africa (Simmons 2021).

A Hybrid Approach for Climate Change Adaptation

The project has successfully implemented a hybrid approach to enhance the project implementation process, fusing the top-down approach government-led and government-implemented and the bottom-up approach, including civil society engagement with PIs at community and local levels (Stochowiak et al., 2016; OECD, 2013; Matland, 1995).

The hybrid development approach is a merger of the government's top-down approach through

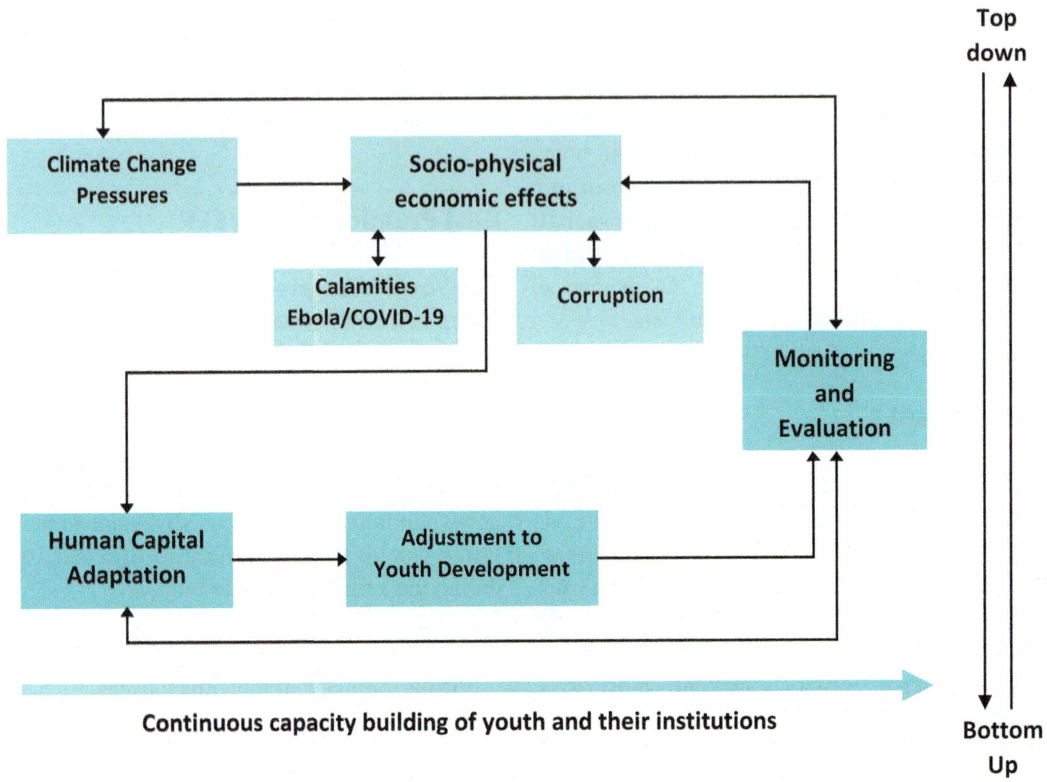

Fig. 6.1 Climate Change Adaptation Framework for youth in entrepreneurship in Africa

the work of the NAYCOM, NCTVA and PPP secretariat in partnership with PIs. NGOs are the main users of the bottom-up approach, including OAS Children Village, CPAD and Global Youth Network, to support the delivery of training to youth in urban and rural communities across Sierra Leone (Matland, 1995; OECD, 2013; Jaja et al., 2016; Stachowiak et al., 2016). The hybrid approach must be viewed as a vital element in the youth development process. It provides opportunities for PIs to work directly with government agencies on issues relating to youth development, employment, entrepreneurship and climate change adaptation (Stochowiak et al., 2016; OECD, 2013). It also provides opportunities for successfully implementing the climate change education for sustainable development models in Africa (UNESCO, 2014, 2015).

NCTVA and PIs, working with the consortium, developed and implemented the curricula utilising the hybrid approach. Government officials and experts from the private sector and civil society agencies, including the Sierra Leone Chamber of Commerce, Industry and Agriculture in Sierra Leone (SLCCASL) and Sierra Leone Institute of Engineers (SLIE), other stakeholders' agencies, participated in the curricula review and approval process. Derived from the hybrid approach, the MOU stands out as a best practice as the single entity that committed the PIs to participate in implementing the project. PIs view the MOUs as a model for validating their contributions to national development planning and sustainable development processes. MOUs were developed and signed by the PIs implementing the project from urban and rural districts in Sierra

Leone. This was intensified by the consultative process implemented during the project's interim stages, which gave rise to the PPP framework. PPP outlines the policy mechanism for the government to work with the business sector within the TVET education subsector. The Government of Sierra Leone, through NAYCOM, increased PI commitment by jointly signing MOUs, which stated that the government would provide tools, technical advice and financial resources to the PIs, while PIs provide their facilities and staff to train the youth and the consortium will provide training materials and finances to support the implementation of the project (OECD, 2013; Poschen, 2015). PIs gained legitimacy and recognition from the government regarding their contribution to achieving the national Sustainable Development Goals.

Active Involvement of PIs in YEEP Development and Recognition

The partner institutions in component 1 were extremely motivated and engaged in the project's overall development to the extent they went beyond their roles and participated in the development of initiatives that add value to the YEEP project. The KE and NKE provided technical advice to senior administrative officials, trainers and instructors of Njala University to develop a project proposal that addressed weaknesses in delivering skills and knowledge to youth attending the Aquaculture and fisheries development faculty. The project proposal was presented to the Youth Commissioner, who presented it to the World Bank for funding, and the US Ambassador of Sierra Leone of the American Embassy with success. The proposal was also submitted to the Ministry of Higher and Technical Education, Government of Sierra Leone, for funding support, and a grant to develop and sustain the aquaculture value chain at Njala University was received.

The second initiative materialised by Njala University was the establishment of the National Aquaculture Association of Sierra Leone. The Aquaculture Department contacted all governmental, private sector and civil society stakeholders involved in Aquaculture production to work with them to create a national Aquaculture Association of Sierra Leone. The Department mobilised seed funding, and the Minister of Agriculture hosted a launching at the Ministry of Agriculture conference room. The National Aquaculture Association of Sierra Leone is now fully functional under the umbrella of the Aquaculture Department at Njala University.

GTI spent its limited resources to provide new spaces and facilities to train youth from adjacent communities in new skills for sustainable livelihoods. The GTI created a separate training facility, using aluminium fabrication, dedicated to delivering the practical courses. Youths undergoing the 1-year certificate course in Sustainable Construction were engaged in a 4-month practical aluminium fabrication training course. MMCET renovated and fitted a building dedicated to training participants in e-travel. It installed new furniture for the e-travel lab that was recommended during the TOT workshop. The operationalisation of the e-travel lab project was delayed, while the institutions mobilise additional financial resources to purchase the necessary computers for the lab.

Increased Participation of Youth

PIs in component 2 increased the number of trained counsellors in the Career Guidance and Job Readiness Skills from 30 counsellors from Freetown, the capital, to 60 counsellors from 7 PIs from districts of Freetown, Mokonde, Bo, Makeni and Kenema. This has led to increased participation of young people in training on Career Guidance and Job Readiness Skills from 2400 to 6130 young people. The Career Guidance and Job Readiness Skills programme effectively provided youth with soft skills in time management and writing business letters, including job applications. This gap in 'finer' skills was viewed as a significant deficiency in the training programmes provided to graduates at tertiary institutions in Sierra Leone (Carter, 2008, ILO, 2010, UNDP, 2015).

PIs Valued as a Contributor to the Achievement of National Development

The institutions, mainly NGOs, participating in the component 3 programme lacked dedicated funding, full-time professional staff and the necessary institutional capacity to run sustained programmes over time. These agencies felt that they are not valued or respected by the government and the donor communities due to the lack of the government's support and funding deficit. The MOU and the provision of financial resources acknowledged that the government values PIs as partners in the national development as they contribute to the national development planning process. This strategy led to the commitment of the PIs. PIs provided their facilities and staff free of charge to the project and increased the numbers of instructors and participants in the project.

Recommendation to Enhance YEEP

Although the piloting of the YEEP was remarkably successful in achieving all the deliverables agreed in the project proposal, there are some notable gaps in the overall management and operation of the project, which need to be addressed to enable the smooth implementation of the upscaling. There is a need to enhance the capacity of the key officials of government to enable them to understand the definitions and the strategic application of important concepts such as climate change stressors and pressures and others. To enhance the implementation of the overall project, some recommendations are provided. The recommendations are divided into two sections. These include (a) specific recommendations to enhance the upscaling of the YEEP and (b) general recommendations to enhance TVET and climate change adaptation initiatives.

Recommendation to Enhance the Upscaling of YEEP in Other African Countries

The following recommendations are provided to support the upscaling and/or implementation of similar climate change adaptation projects for youth in entrepreneurship in Africa:

- The strengthening of the relationship between partner institutions engaged in the post-implementation of YEEP at the national and regional levels. This can be done by maintaining the consultative forum and outreach extension activities to partner institutions by sharing information on the performance of the training courses and other project activities. The consultative forums will ensure the full participation of all partner institutions in implementing the YEEP and further youth development projects implemented by the governmental, IGOs, NGOs and other development agencies.
- The implementing agency should ensure the continuous monitoring of the curricula in all partner institutions to ensure that the training links to the skills required by the businesses sector and the market. The role can be conferred to the National Youth Commission, the Ministry of Youth Districts and their appropriate agencies working at district levels in collaboration with the M&E manager at the national level. The work should be supervised by a special committee involving representatives of business agencies, the NCTVA, the Ministry of Labour, the Ministry of Higher Education and other relevant stakeholders. Data on the training appropriateness, relevance and quality should be sent to the Ministry of Labour and Chamber of Commerce, Industry and Agriculture for their feedback and follow-up.
- The implementing agency should ensure that the person/s employed as project manager/s

possesses critical project management and people skills; the lack of skilled expertise in these areas of the YEEP project manager created unnecessary challenges for the consortium staff to pilot the project.

- The relationship between the implementing agency and the private sector was weak; the implementing agency needs to work urgently to strengthen such relationships by establishing programmes to facilitate effective collaboration and partnership between critical governmental and the private sector agencies. This model will lead to the development of effective partnerships with key stakeholders involved in the TVET subsector.
- In the absence of a worthwhile revolving national youth development fund to support youth entrepreneurs to develop and finance their project, the implementing agency should work with the Ministry of Youth and the Ministry of Finance to create a Micro Credit and Grant funds facility for the disposition of the Commonwealth Youth Credit Initiative. The Micro Credit Facility is a revolving loan and grants fund which provides funding to YEEP graduates who have viable business ideas or business plans ready for implementation. With such seed capital supplied to them, it will create a new zeal for undergraduates at universities, colleges and TVET institutions to nurse viable business ideas with the opportunity to bring their vision into reality through YEEP.
- The implementing agency should invest in online platforms and applications facilities that facilitate the growth and expansion of entrepreneurship opportunities and collaboration among entrepreneurs, coaches, financiers and partner institutions to move the YEEP agenda forward.
- Due to the benefits derived from the pilot project, the implementing agency and the consortium should put measures in place to ensure the upscaling of the YEEP pilot project to a national project addressing the critical issues raised during the implementation of the project.

Recommendations to Enhance Overall TVET

- Stimulate the creation of formal and sustainable green jobs in countries where high shares of informal employment form a significant obstacle to better social mobility. Therefore, economic progress policies should create more enterprises in the formal sector that offer standard decent green jobs. This approach can be addressed by economic policy reforms such as abolishing bureaucratic business registration procedures and tax reforms, stimulating investment in the private sector, investing in appropriate and relevant TVET training and creating formal companies' start-up support.
- Modernise TVET centres by strengthening the vocational part of their educational or schooling system and bring existing vocational education and training systems closer to the current needs of the labour market so that young people can experience a smoother transition to jobs. Vocational education should be updated and accompanied with phases of practical work experience via internships and/or apprenticeship attachment. Employers should be consulted regarding the design and delivery of TVET curricula.
- Bring academic education closer to the private sector in the situation of the YEEP, i.e. university graduates who are experiencing difficulties finding jobs, to make academic training a more labour market-oriented process. It is necessary to integrate internships with employers into educational curricula to acquire knowledge and skills with work practices in the private sector. This approach would require universities and colleges to modify academic curricula to reflect the needs of the market.
- Promote and enhance the quality of TVET programmes in situations where there are low

levels of acceptance among the public who perceive vocational training as an inferior type of professional education. In addressing this situation, it is essential to develop and implement programmes that raise the awareness of youth and their parents regarding the opportunities and benefits of pursuing TVET education streams. The government should provide resources to enhance the quality of the TVET initiative to create decent green jobs.

- Implement empirical research to generate better data on TVET and the application of Climate Change Education for Sustainable Development. Research on the value of vocational training would benefit enormously from better data and a suitable programme design to evaluate policy initiatives. M&E systems should be established in TVET programmes to ensure the universal collection of assessment data.
- Upgrade vocational training in the informal sector to provide better training for the informal sector. There should be a shift from the traditional or casual apprenticeship system that is limited to crafts and other conventional products and skills, linking the training to the school system and the formal sector, including new technologies and occupational change. This notion requires the government to recognise informal employment as part of the economic and social reality to drive sustainable development goals.

Conclusion

Countries in West Africa struggle with finding sustainable employment for the expanding armies of young people, who are constantly steering into a perilous bleak future plagued with unemployment, poverty and under-development. Policymakers, development practitioners and planners should collaborate with donor agencies and the international community to find workable, practical solutions to generate green and decent jobs. The Youth Employment and Entrepreneurship Project (YEEP) demonstrates that the Climate Change Adaptation Framework on youth entrepreneurship model possesses the best possible solution to address Africa's unemployment and poverty situation. The project showed that it is highly probable to resolve the mismatch between the skills and knowledge provided by education and training institutions required by the job market while moving forward in building a sustainable green economy in the future. The road ahead is particularly challenging and uncertain as the country continues to grapple with development calamities such as COVID-19, Ebola and other pestilences.

McGarth and colleagues (2020) argue that credible, specific skills systems can be built to support individuals, communities and organisations to produce and sustain improved livelihood opportunities in Africa. However, the primary current theoretical approaches to TVET are not up to this challenge.

The project demonstrates that as youth proceed on a journey to a sustainable future where they actively shape their future, positive outcomes are possible. However, these outcomes are viable when there is a comprehensive understanding of the challenges facing people and opportunities to address these development challenges. It is imperative to bear in mind that there is no quick fix or instant coffee approach to achieve sustainable development at global levels. People are required to work collectively by putting their shoulders to the wheel and make the necessary sacrifices to bring about the changes required to build the resilience of youth to adapt to the impact of climate change with emphasis on resolving biodiversity and ecosystems degradation of land, sky and sea.

References

Angelsen, A., Larsen, H. O., Lund, J. F., Smith-Hall, C., & Wunder, S. (2011). *Measuring livelihoods and environmental dependence; methods for research and fieldwork*. Earthscan.

Bernick, C. L. (2001). When your culture needs a makeover. *Harvard Business Review*, 53–61.

Carter, R. (2008). *Caribbean youth: An integrated literature review*. Bridgetown.

Drucker, P. (2006). *Classic Drucker: Essential wisdom of Peter Drucker from the pages of Harvard business review*. Harvard Business Press.

Earl, M. J. (2001). IT strategy in the new economy. In J. Pickford (Ed.), *Mastering management 2.0: Your single-source guide to becoming a master of management, financial times*. Prentice Hall Publishing.

Government of Sierra Leone. (2019). *The Government of Sierra Leone's new Medium-term National Development Plan (MTNDP) 2019–2023*. Freetown.

Hamel, G. (2001). *Leading the revolution*. Harvard Business School Press.

Handy, C. (1986). *Understanding organization* (4th ed.). Penguin Books.

Harley, M., & van Minnen, J. (2010). *Adaptation indicators for biodiversity*. European Topic Centre on Air and Climate Change Technical Paper 2010/15, European Environment Agency.

Hernandez-Delgado, E. A. (2015). *The emerging threats of climate change on tropical coastal ecosystem service, public health, local economies and livelihoods sustainability of small islands: Cumulative impacts and synergies*. Marine Pollution Bulletin, University of Puerto Rico, Centre for Applies Tropical Ecology and Conservation, Coral Reef Research Group Puerto Rico, USA.

Jaja, J., Dawson, J., & Gaudet, J. (2016). *Using social network analysis to examine the role institutional integration plays in community-based adaptive capacity to climate change in Caribbean small island communities*. Department of Geography, Environment and Geomatics, University of Ottawa, Canada: Routledge, Local Environment.

Johnson, G., & Scholes, K. (2002). *Exploring corporate strategy* (6th ed.). Prentice-Hall.

Kanyandago, P. (2010). Revaluing the African endogenous education system for community-based learning. In J. Zeelen, J. van der Linden, D. Nampota, & M. Ngabirano (Eds.), *The burden of educational exclusion* (pp. 99–114). Rotterdam, Boston and Taipei.

ILO. (2010, November). *A skilled workforce for sustainable growth: A G20 training strategy*. International Labour Office.

IMF. (2019, July). *Sierra Leone: Economic development documents-national development plan, 2019–23* (Vol. 2019/218). International Monetary Fund. African Dept..

IPCC. (2014a). Climate change (2014): Synthesis report. In *Contribution of working groups I, II and III to the fifth assessment report of the intergovernmental panel on climate change* (p. 151). IPCC.

IPCC. (2014b). Summary for policymakers. In: Climate change impacts, adaptation and vulnerability. In C. B. Barros, V. R. Dokken, D. Mach, K. J. Mastrandrea, M. D. Bilir, T. E. Chatterjee, M. Ebi, & K. L. Estrada (Eds.), *Contributions of the Working Group II to the Fifth Assessment Report*. Field.

IPCC. (2014c). *Climate change: Synthesis report. In: Contribution of working group 1, 11, 111 to the fifth assessment report of the intergovernmental panel on climate change* (p. 151). IPCC.

McGrath, S., Ramsarup, P., Zeelen, J., Wedekind, V., Allais, S., Lotz-Sisitka, H., Monk, D., Openjuru, G., & Russon, J.-A. (2020). Vocational education and training for African development: A literature review. *Journal of Vocational Education and Training, 72*(e4), 465–487.

Matland, R. (1995). Synthesising the implementation literature: The ambiguity-conflict model of policy implementation. *Journal of Public Administration Research and Theory, 5*(2), 145–174.

Nurse, L. A., McLean, R. F., Agard, J., Briguglio, L. P., Duvat-Magnan, V., Pelesikoti, N., Tompkins, E., & Webb, A. (2014). Small islands. In V. R. Barros, C. B. Field, D. J. Dokken, M. D. Mastrandrea, K. J. Mach, T. E. Bilir, M. Chatterjee, K. L. Ebi, Y. O. Estrada, R. C. Genova, B. Girma, E. S. Kissel, A. N. Levy, S. MacCracken, P. R. Mastrandrea, & L. L. White (Eds.), *Climate Change 2014: Impacts, Adaptation, and Vulnerability. Part B: Regional Aspects*. Contribution of Working Group II to the Fifth Assessment Report of the Intergovernmental Panel on Climate Change Cambridge University Press, Cambridge, United Kingdom and New York, NY, USA, pp. 1613–54.

Niyonkuru, F. (2016). Failure of foreign aid in developing countries: A quest for alternatives: Business and economics. *Business and Economics Journal, 2016*(7), 3. https://doi.org/10.4172/2151-6219.1000231

OECD. (2013). *The nature of policy change and implementation: A review of different theoretical approaches*. Organisation for Economic Cooperation and Development.

Poschen, P. (2015). *Decent work, green jobs and the sustainable economy: Solution for climate change development*. International Labour Organisation.

Pereznieto, P., Gbedemah, C., Rosenem, G., Haper, C., & Jones, N. (2011). *Youth vulnerability and adaptation: Exploring the impact of macro-level shocks on youth 3F crisis and climate changes*. Overseas Development Institution.

Prahalad, C. K. (2006). *The fortune at the bottom of the pyramid; eradicating poverty through profits; enabling*. Warthons School Publishing, Pearson Education.

Schindler, M., & Eppler, M. (2003). Harvesting project knowledge: A review of project learning methods and success factors. *International Journal of Project Management, 21*(3), 219–228. https://doi.org/10.1016/S0263-7863(02)00096-0

Schleussner, C. F., Lissen, T. K., Fischer, E. M., Wohland, J., Perrette, M., Golly, A., Rogelj, J., Childers, K., Schewe, J., Frieler, K., Mengel, M., Hare, W., & Schaeffer, M. (2016). Different climate impact for policy-relevant limits to global warming: The case of 1.5 °C and 2 °C. *Earth System Dynamics, 7*, 327–351.

Simmons, A. (2021). *Impact of climate change on Youth in small island communities* (Sustainable development goals series, climate change) (Vol. 1). Springer Nature.

Slack, N., Chambers, S., Harland, C., Harrison, A., & Johnston, R. (1998). *Operations Management* (2nd ed.). Pearson Education Limited.

Stochowiak, S., Robles, L., Habtermanion, E., & Maltry, M. (2016). *Beyond win: Pathway for policy implementation*. ORS Impact/The Atlas Learning Project.

Thuraya, K. (2015). *Improving the effectiveness of E-learning improvement in the school of engineering at Tripoli University*; Doctoral thesis, University of Huddersfield.

Tukundane, C., Minnaert, A., Zeelen, J., & Kanyandago, P. (2015). Building vocational skills for marginalised youth in Uganda. *International Journal of Educational Development, 40*, 134–144.

UNDP. (2019). *Sierra Leone 4th National Human Development Report 2019: Building Resilience for Sustainable Development*. United Nations Development Programme.

UNDP. (2015). *Human development report*. Work for Human Development, United National Development Programme (UNDP).

UNDP. (2021). *GEO-6 for Youth*. UN Environment Programme.

UNESCO. (2014). *Shaping the future we want: UN decade of education for sustainable development (2004–2014)*. DESD Monitoring and Evaluation.

UNESCO. (2015). *Not just hot air: Putting climate change education into practice*. Education sector.

Vignos, T. (2014). *Implementing lessons learned best practices in project management*. Capstone Report; Enterprise Project Manager Pitney Bowes, Inc.; University of Oregon Applied Information Management Program; University of Oregon Eugene.

Appendices

Appendix I Questions to Be Used During Focus Group Discussions with Youth

1 ID

TVET Centre, BSCs, Colleges and Universities

Collecting data for YEEP M&E

FGD with youth attending courses at PIs

Questions for Focus Group Discussion (4 students per PIs due to COVID-19 pandemic)

Please fill in the information below

Names of youth: ______________________________
____
____

Job title:
____
____

Age range: 15–19 ( ), 20–24 ( ), 25–29 ( ), 30–34 ( ), 35 +( ).

Address: ______________________________
____
____

Contacts: Phone ____________ Email: ___
____

Names of partner institution: ____________
____
____

Name of course attended:
____
____

A. Simmons, *Climate Change Adaptation Framework and Youth Entrepreneurship in West Africa*, Sustainable Development Goals Series, https://doi.org/10.1007/978-3-030-85754-7

Please answer the questions below

Why are you attending the training programme?

1. What do you like most about the training project? Why?
2. What do you dislike most about the training project? Why?
3. What is your impression of the methods used by the instructors/counsellors to deliver the training to youth?
4. Do you think the methods used by the instructor and counsellors to deliver the training course are effective? Explain.
5. If you look at the whole training programme, a) what aspects of training courses have been working well? And b) what aspects of the training programme have not been working well? Please explain.
6. What do you think can be done to improve the delivery of the training?
7. Would you recommend such training to be provided in the future? Why?
8. What are the key learnings that you can take away from the YEEP training programme which would impact on the career and employment of youth in the future?

Appendix II Questions Used During Interviews with Policymakers and Senior Officials

2 *ID*							Collecting data for YEEP M&E
TVET Centre, BSCs, Colleges and Universities							Interviews with Heads of PIs and Instructors

Questions for Key Informant (2 members of staff per institution)

Please fill in the information below

Name of official/staff

____

____

Job title:

____

____

Age range: 20–24 ( ), 25–29 ( ), 30–34 ( ), 35–39 ( ), 40 + ( ) Please Ticket

Address:

____

____

Contact: Phone: ____________________ Email:

____

Names of partner institution:

____

____

Name of course taught in your institution:

____

____

Interview Administered by YEEP staff (Project Leader)

Please respond to the following questions

1. Tell us a bit about yourself and why you and your institution decide to participate in this project?
2. What do you see are the strengths of this project?
3. What are the identified challenges facing the project?
4. What are the methodologies used for training youth participating in the project? Please explain.
5. How effective are these methods?
6. What is the relevance of this project to the overall development of youth employment and sustainable livelihoods in Sierra Leone?
7. What are the key lessons learned from this project in terms of moving forward toward youth employment and career development?
8. What recommendations would you consider for improving the YEEP and similar youth development projects?
9. What can be done to upscale the project from a pilot to a national youth development and employment project?

Appendix III Questions to Enable Consortium Officials to Prepare Case Study on Youth

3 ID

TVET Centre, BSCs, Colleges and Universities

Collecting data for YEEP M&E

For the preparation of Youth case studies from components 1, 2 & 3

Case Studies (2 persons per institution)

Please fill in the information below

Name of person: ______________________________

Age range: 15–19 ( ), 20–24 ( ), 25–29 ( ), 30–34 ( ), 35 +( ). Please think the appropriate box

Address: ______________________________

Contacts: Phone ______________ Email: ______________________________

Names of partner institution attended: ______________________________

Name of course attended: ______________________________

Please answer the questions below.

Why did you decide to participate in the project? ______________________________

What are the key gains of the project for you? Please explain. ______________________________

What are the main strengths of this programme? Please explain.

____
____
____
____
____
____
____

What are the shortcomings/weaknesses of the programme? Please explain.

____
____
____
____
____
____
____

What are you doing to address these shortcomings/weaknesses in the training programme?

____
____
____
____
____
____
____

What lessons would you extract from this programme that will help you with future employment and your career?

____
____
____
____
____
____
____

What recommendations can you give to improve the project?

____
____
____
____
____
____

Appendix IV Questions Used During the Administering of Questionnaires with Heads of Institutions and Instructors of Partner Institutions

4 ID

Policy Makers and Senior Officials of the Government of Sierra leone

Collecting data for YEEP M&E

Interviews with Policy Makers and Senior Officials

Questions for Key Informant (2 persons per institution)

Please fill in the information below.

i) Name of official: ____ ____ ____

ii) Job title of the official: ____ ____ ____

iii) Age range: 20–24 ( ), 25–29 ( ), 30–34 ( ), 35–39 ( ), 40 + ( ) Please Tickct

iv) Address: ____ ____ ____

v) Contacts:Phone:____________________ E m a i l : ______________________________

vi) Names of partner institution: ____ ____ ____

vii) Name of course: ____ ____ ____

Interview Administered by Project Lead
Please answer the questions below.

1. Tell us a bit about yourself and why you decide to be part of this programme?
2. What do you see as the strengths of the YEEP?
3. What are the challenges do you see facing the project?
4. What are the methodologies used by instructors and counsellors to deliver the training to the youth?
5. How effective are the training methods used by instructors and counsellors when delivering the project?
6. What is the relevance of this project to the government's overall thrust to develop the youth of Sierra Leone, specifically in employment and career development?
7. What do you see are the lessons learned from the project regarding the areas of youth employment and career development?

8. What recommendations do you wish to put forward for consideration to improve the YEEP and similar youth development projects in the future?
9. What can be done to up-scale the project from a pilot project into a national youth development project?

Index

A. Simmons, *Climate Change Adaptation Framework and Youth Entrepreneurship in West Africa*, Sustainable Development Goals Series, https://doi.org/10.1007/978-3-030-85754-7

The manufacturer's authorised representative in the EU is Springer Nature Customer Service Centre GmbH, Europaplatz 3, 69115 Heidelberg, Germany. If you have any concerns regarding our products, please contact ProductSafety@springernature.com

Printed and bound by CPI Group (UK) Ltd, Croydon, CR0 4YY
28/09/2026
02224079-0008